Forensic Investigative Genetic Genealogy (FIGG) and DNA Evidence

Forensic Investigative Genetic Genealogy (FIGG) and DNA Evidence is a vital guide for professionals and practitioners, providing the most complete and current view of the use of DNA from crime scenes and its comparison with DNA test results available on online databases.

Across nine chapters, and through a practical lens, the book highlights how FIGG can generate investigative leads and identify potential pools of suspects. The book addresses key issues in modern forensic genealogy, such as the emergence of AI in the forensic field, as well as how forensic genetics deals with the application of genetic knowledge as applied to legal problems and litigation. Finally, the book will also examine law enforcement agencies and trained forensic genetic genealogists who have developed trees and generated leads in family trees to solve multiple cold cases.

With contributing chapters from leading experts in the field, *Forensic Investigative Genetic Genealogy (FIGG) and DNA Evidence* offers invaluable insights into the evolving landscape of forensic genetics and genealogy usage in criminal and cold case investigations. It is an essential resource for professionals seeking to understand the current capabilities and future possibilities of these groundbreaking investigative tools.

Forensic Investigative Genetic Genealogy (FIGG) and DNA Evidence

A Guide for Practitioners

Edited by
John R. Vacca

CRC Press
Taylor & Francis Group
Boca Raton London New York

CRC Press is an imprint of the
Taylor & Francis Group, an **informa** business

Designed cover image: Getty Images (artacet)

First edition published 2026
by CRC Press
2385 NW Executive Center Drive, Suite 320, Boca Raton FL 33431

and by CRC Press
4 Park Square, Milton Park, Abingdon, Oxon, OX14 4RN

CRC Press is an imprint of Taylor & Francis Group, LLC

ISBN: 978-1-032-71497-4 (hbk)
ISBN: 978-1-032-71498-1 (pbk)
ISBN: 978-1-032-71499-8 (ebk)

DOI: 10.4324/9781032714998

Typeset in Sabon
by SPi Technologies India Pvt Ltd (Straive)

In memory of Joseph Dosio (1940–2009)

Contents

About the Editor

John R. Vacca is an information technology consultant, researcher, professional writer, editor, reviewer, and internationally known, best-selling author based in Pomeroy, Ohio, United States. Since 1982, John has authored, edited, and published more than 87 books (some of his most recent published books include):

- **Cloud Computing Security: Foundations and Challenges, 2nd Edition** (Publisher: CRC Press (an imprint of Taylor & Francis Group, LLC (November 2, 2020))
- **Security in the Private Cloud** (Publisher: CRC Press (an imprint of Taylor & Francis Group, LLC (September 1, 2016))
- **Cloud Computing Security: Foundations and Challenges** (Publisher: CRC Press (an imprint of Taylor & Francis Group, LLC (August 19, 2016))
- **Handbook of Sensor Networking: Advanced Technologies and Applications** (Publisher: CRC Press (an imprint of Taylor & Francis Group, LLC (January 14, 2015))

as well as more than 600 articles in the areas of advanced storage, computer security, and aerospace technology (copies of articles and books are available upon request).

John was also a configuration management specialist, computer specialist, and the computer security official (CSO) for NASA's space station program (Freedom) and the International Space Station Program, from 1988 until his retirement from NASA in 1995.

In addition, John is also an independent online book reviewer. Finally, John was one of the security consultants for the MGM movie titled: "AntiTrust," which was released on January 12, 2001. A detailed copy of John's author bio can be viewed at URL: https://www.johnvacca.com. John can be reached at: john2164@windstream.net

Contributors

Richard Allen Anderson
Department of Humanities & Social Sciences
G41 Project
Fargo, USA

Heather Miller Coyle
Department of Humanities & Social Sciences
University of New Haven
West Haven, USA

Tracey Leigh Dowdeswell
Douglas College
New Westminster, Canada

Steven Smugeresky
Montgomery County Department of Police
Gathersburg, USA

Nicoladie Tam
Department of Humanities & Social Sciences
University of North Texas
Denton, USA

Preface

SCOPE OF COVERAGE

This comprehensive edited handbook serves as a professional reference, as well as a practitioner's guide to today's most complete and concise view of the use of DNA (deoxyribonucleic acid) from crime scenes and comparing it with the DNA test results available on online databases. The focus will be on forensic investigative genetic genealogy (FIGG) and the GEDmatch database, as it is an online database being used to solve cold cases, though it will touch on the CODIS database, which is network accessible only on a limited basis—through designated terminals at approved forensic crime labs. It offers an in-depth coverage of how FIGG can be used to provide other investigative leads and present a potential pool of suspects. A functional approach will be adopted regarding the classification of the use of forensic genealogy on the most difficult cases, which would help to unlock investigative leads. Most labs using FIGG use a triage system, and some states require the lab to exhaust all other leads and methods, seeking approval from a judge prior to proceeding. The book will address such issues.

The primary audience for this handbook consists of international law enforcement professionals and criminal justice practitioners, law enforcement officers, and academics, who share experiences and ideas on innovative developments in any aspect of forensic genetics. This edited handbook is also designed to introduce professional and public audiences to the use of DNA in criminal investigations; to understand what DNA can and can't tell about a crime, and what the current and future uses of DNA analysis in the criminal justice system might be. Forensic genetics DNA analysis is a complex area open to misinterpretation. So, the contributors will have set out to provide a straightforward edited handbook for forensic genetics DNA analysts, police, judiciary, lawyers, jurors, journalists, and those who are intrigued by criminal casework. In other words, this would be anyone with an interest in the use of DNA for crime investigation.

ORGANIZATION OF THIS BOOK

The book is organized into six sections composed of 9 contributed chapters by leading experts in their fields; as well as five appendices, including an extensive glossary of forensic genetics terms and acronyms.

Chapter 1, "Introduction to Forensic Genetics," sets the stage for the rest of the book by providing a comprehensive examination of the foundational principles and cutting-edge advancements in forensic genetics and highlighting its vital role within the criminal justice system. In addition, it opens with a clear introduction to the core concepts and historical timeline of forensic genetics.

Chapter 2, "How DNA Profiling Is Used in the Criminal Justice System," provides a detailed description of the multifaceted role of DNA profiling within the criminal justice system by examining its development, implementation, and associated legal and ethical concerns. Beginning with a survey of the origins and evolution of DNA profiling, the chapter outlines key milestones in its adoption and discusses the legislative frameworks that govern its use.

Chapter 3, "Biogeographic Ancestry Testing," explores the principles and methods of BAT, its use in criminal justice, and the legal and ethical issues surrounding its use, especially concerning civilian genetic databases. The different viewpoints of law enforcement and defense advocates are examined, including legal precedents and current challenges, to provide a comprehensive understanding of BAT's role in modern forensics and the criminal justice system.

Chapter 4, "The Accuracy of Forensic DNA Testing," describes step-by-step the basic forensic DNA test methods used in standard human identification and discusses the accuracy of the major methods and decision points used in this process. Included in this chapter are factors that influence accuracy; evidence collection sterility; the evidence sampling strategy for testing; the choice of DNA extraction and quantitation methods; PCR amplification efficiency; instrument sampling for electrokinetic injection; and software analysis settings. Specialty DNA methods such as mitochondrial DNA sequencing, YSTRs, and SNP testing are also discussed in this chapter.

Chapter 5, "Why Forensic Genealogy Is Important," explores the brief case studies of solved FIGG cases, and the summarized reflections of persons involved (and the human impact) to illustrate the importance of FIGG.

Chapter 6, "FIGG and How It Is Useful in Solving Crimes," discusses the process from start to finish of a FIGG case and the best practices an investigator should review when utilizing this investigative discipline.

Chapter 7, "Recent and Prominent Cases Solved by Forensic Genetics Science Utility Technology," focuses on cases that are prominent for what they can teach practitioners about the development of the field, particularly the ideas and practices that are pushing its limits, expanding its capabilities, and breaking new ground in forensic investigations.

Chapter 8, "Ethics as Lived Practice: Anticipatory Capacity and Ethical Decision-Making in Forensic Genetics," focuses on emerging norms as embodied in professional codes of ethics, practice standards, and policies that are being developed to guide the ethical practitioner of FIGG.

Chapter 9, "Artificial Intelligence and Genealogy," explores quantitative genealogical and forensic analyses of genetic data, illustrated with case examples that introduce the core principles of probabilistic inheritance and analytical techniques for determining ancestral lineages and identity matches. The discussion emphasizes the use of short DNA polymorphism variant sequences, including single-nucleotide polymorphisms (SNPs), short tandem repeats (STRs), and microhaplotypes (MHs), as biomarkers. Forensic DNA index system databases were also discussed.

John R. Vacca
Managing and Consulting Editor
TechWrite
Pomeroy, Ohio

Acknowledgments

There are many people whose efforts on this book have contributed to its successful completion. I owe each of you a debt of gratitude and want to take this opportunity to offer my sincere thanks.

A very special thanks to my Senior Editor, Mark Listewnik, without whose continued interest and support this work would not have made this book possible; and to Editorial Assistants, Ella Halstead and Niamh Hitchmough, who provided staunch support and encouragement when it was most needed. I am also grateful to my Project Editor, Jayapriya Thillaikkarasu; and the copyeditors and proofreaders, whose fine editorial work has been invaluable. Finally, thanks to all of the other people at CRC Press/Taylor & Francis Group, whose many talents and skills are essential to a finished book.

Thanks to my wife, Mary Vacca, for her love, her help, and her understanding of my long work hours. Finally, I wish to thank all the following authors who contributed chapters that were necessary for the completion of this book: Rick Anderson, Heather Miller Coyle, Steven Smugeresky, Tracey Leigh Dowdeswell, and Nicoladie Tam

CHAPTER 1

Introduction to Forensic Genetics

Richard Allen Anderson

CONTENTS

1.1 INTRODUCTION

In the rapidly evolving landscape of forensic science, genetic technology and forensic investigations have converged to pursue justice [1–3]. As society grapples with complex and controversial challenges at the intersection of technological advancements, forensic genetics, public safety, justice, and legal issues, forensic genetics in the criminal justice system has emerged as a scientific discipline of profound consequence [1, 3–8]. This chapter explores the history, foundation, advancements, and legal implications of DNA forensic genetics. It presents the pioneering journey of DNA forensic science, which began in the 1980s [9], and its present-day role in the criminal justice system and the constitutional rights of individuals [1, 10–12].

DOI: 10.4324/9781032714998-1

1.1.1 The Intersection of Genetics and Justice

Forensic genetics is a dynamic fusion of molecular biology and criminal investigation in the pursuit of justice for victims, offenders, and the falsely accused alike [13–18]. Once viewed as the holy grail for the prosecution, DNA forensic science has also brought justice for many falsely convicted of crimes they did not commit by exposing prejudicial, incompetent, malicious, and even intentionally corrupted criminal investigations and prosecutions, finally bringing justice to many of the innocent whose lives were destroyed by decades of imprisonment for crimes they did not commit [1, 6, 7, 15, 16, 19–23]. Since its inception in the 1980s, DNA forensic science, specifically in the context of the criminal justice system (which is the focus of this book), has evolved rapidly due to the catalysts of technological advancements in DNA sequencing, bioinformatics, and polymerase chain reaction (PCR) technologies [2, 4, 24–28].

The view of DNA forensic science as the supposed 'holy grail' for prosecution in the criminal justice system has blinded many to the vulnerabilities in the collection, preservation, and analysis of DNA evidence, and has traditionally left many underinformed criminal defense counsel at a loss not fully understanding how to construct a valid defense using DNA forensics, for that percentage of their clients who *may* actually be falsely accused [1, 7, 8, 13, 17, 18, 20, 22]. *Reasonable doubt* certainly still applies in a court of law, considering historical issues, legal challenges, and the prospect of any number of circumstances from the initial investigation of a crime scene through the sequence of DNA analysis and reporting, now more than ever, criminal defense has the tools to build arguments that cause *reasonable doubt* in the minds of judges and jurors [1, 2, 4, 6, 13, 15, 18, 20–22, 28–39].

1.2 DEFINING FORENSIC GENETICS

At the core of DNA forensic genetics lie its foundational principles, methodologies, and legal issues, and in many cases, constitutional ones [2, 4, 27, 29]. The primary focus of DNA forensic science is the analysis of biological samples, including blood, hair, bone material, saliva, semen, and other bodily fluids and materials. Analysis of focused regions of the genome has enabled forensic scientists to generate DNA profiles for comparison to those of suspects, victims, relatives, and missing persons. This allows, with high degrees of certainty, the identification or exclusion of individuals' connection to a crime, or solving cold cases from decades, even centuries beforehand [24, 28, 40]. Forensic genetics also plays a role in civil cases, including paternity or kinship analysis, in disaster victim identification (DVI) following mass fatality incidents [28, 32].

Cold and missing persons cases can now often be resolved through DNA analysis when unidentified remains are discovered [24, 41, 42]. In many instances, this interrelates with cases of wrongful convictions later exonerated because of DNA evidence, [1, 7, 8, 20, 22, 43, 44], this this author asserts the close relationship between cold cases and wrongful convictions, and wherever an inkling of doubt or consistent defense claim of innocence exists, it should be the public duty, in the interest of justice, to reexamine DNA in older cases where convictions were found before the common use of DNA. Using DNA along with other forensic evidence may also determine whether foul play or accidental causes led to an individual's demise, which can result in opening cold cases and, in some instances, leading to the conviction of guilty parties or the freeing of those wrongfully convicted from decades-old cases.

1.2.1 Historical Evolution and Impact in Criminal Justice

The emergence of DNA profiling in the 1980s marked a paradigm shift in criminal investigations. Unlike traditional methods of identification, such as fingerprinting or eyewitness testimony, DNA evidence provides a level of specificity unparalleled in forensic science. The probability of two unrelated individuals sharing the same DNA profile, except for identical twins, is exceedingly low, making DNA a powerful tool in judicial proceedings [2, 24, 34]. Its reliability, however, hinges on rigorous laboratory protocols, robust interpretation standards, and strict chain-of-custody procedures [9, 28, 34, 40, 45].

As DNA profiling emerged in the 1980s, this marked a paradigm shift in criminal investigations. The use of DNA evidence has provided a more technical and sophisticated level of forensic science, surpassing traditional methods such as fingerprinting or eyewitness identification. Eyewitness identification has known flaws and vulnerabilities [2, 6, 8, 10, 18, 20, 34–37, 40, 46, 47]; however, the possibility that two *unrelated* individuals share the same DNA is either unlikely or extremely rare. Unlike earlier identification methods—such as fingerprints or serology—DNA profiling provides an extraordinarily higher level of accuracy. With billions of possible combinations at the analyzed loci, the chance of two unrelated individuals sharing the same DNA profile is astronomically low (often cited as less than one in a billion, depending on the number of loci tested) [48–50].

As one discovers the benefits of DNA analysis in the criminal justice system, it is essential to consider that false confidence has traditionally been given to the use of DNA evidence in criminal cases. There are many instances of mistakes, shortcuts, and manipulation or tampering (intentional or accidental) that have resulted in wrongful convictions [1, 6, 8, 16, 20–22, 35–37, 51–54].

These instances of corruption have contributed to innumerable wrongful convictions in recent years, shedding a serious negative light on the integrity of law enforcement itself and its forensic scientists working in crime labs. To assume that law enforcement is beyond reproach, considering the mounting number of cases that prove otherwise, can be detrimental to those who may be falsely accused. Law enforcement officers are just as human as the criminals they arrest, accuse, and aid in prosecuting. Considering the relatively low national standards for becoming a law enforcement officer, such as only requiring a GED and no felony convictions (according to many law enforcement agencies' hiring policies still today), society cannot truly expect too much from law enforcement officers in general. Unfortunately, most of society elevates them to an unrealistic, super-human status, not questioning either their intelligence or integrity, when in reality, it should be questioned on every report, in every criminal case brought, to determine whether the officer has not only the training, education, and experience to be a credible witness for the prosecution, but the integrity, the lack of bias, to be impartial [55].

There is an unfortunately high number of law enforcement officers who compromise their integrity on almost a daily basis by targeting certain 'repeat offenders" in their community, stalking bars close to closing time waiting for a certain citizens' vehicle to leave after muttering "so and so if going to get a DUI tonight"—waiting for so and so to leave the bar, following them and doing their traffic stop! this author has witnessed this in my law enforcement career, and though he reported this, the department head refused to address the situation. Corruption exists in the ranks, and bias is often seen as a badge of "experience." Such behaviors need to be thoroughly addressed by defense attorneys and disciplinary committees. These "seeds" of misconduct fester and grow into all forms of corruption in the criminal justice system. Thus, defense advocates should be aware, and diligent to question law enforcement on every level of professional and personal conduct, and look for potential biases individual officers may hold through patterns of behavior, comments made in public or private, and other prospective evidence to question their integrity—to ensure their actions and testimony are not tainted and to provide the best defense to their clients. Furthermore, defense advocates need to educate themselves and utilize independent forensic laboratories to reanalyze evidence, conducting due diligence to prevent, as much as possible, the atrocities of law enforcement and prosecutorial corruption, as well as wrongful convictions [1, 7, 20–23].

1.3 CORE TECHNIQUES AND TECHNOLOGIES IN FORENSIC GENETICS

Many successes in forensic genetic science are attributed to techniques and technologies that have significantly shaped the industry. These include Short Tandem Repeat Analysis, Polymerase Chain Reaction, Mitochondrial DNA Analysis, Y-Chromosome Analysis, Next-Generation Sequencing, and Forensic Investigative Genetic Genealogy. Forensic genetic science has been an ever-evolving field. Considering the rapid development of new and advanced technologies and techniques over the past two to three decades, one can anticipate significant potential for continued advancements in this field in the future. Present techniques and technologies are shown below:

- Short Tandem Repeat (STR) Analysis: STRs are repetitive DNA sequences scattered throughout the genome. By analyzing the number of repeats at specific STR loci, forensic scientists generate a unique DNA profile for each individual. Modern forensic kits, such as the FBI's Combined DNA Index System (CODIS), typically use 20 or more STR markers, providing a powerful tool for human identification [28, 32].
- Polymerase Chain Reaction (PCR): PCR revolutionized forensic genetics by enabling the amplification of minute or degraded DNA samples. This technique makes it possible to analyze even trace amounts of biological evidence [1].
- Mitochondrial DNA (mtDNA) Analysis: mtDNA, inherited maternally, is particularly valuable for analyzing degraded specimens, such as ancient remains or hair shafts without roots. Although less discriminating than STR analysis, mtDNA can be crucial in missing persons and historical investigations [28, 34].
- Y-Chromosome Analysis: The Y-chromosome is passed from father to son and can help in cases involving male lineage, such as sexual assault cases with male perpetrators [28].
- Next-Generation Sequencing (NGS): NGS technologies allow for the simultaneous analysis of thousands of genetic markers, including SNPs (single nucleotide polymorphisms), which can enhance the identification of individuals and provide information about ancestry, physical appearance, and even certain health conditions [32, 46].
- Forensic Investigative Genetic Genealogy (FIGG): FIGG has gained prominence in recent years, allowing law enforcement to solve cold cases using public genealogy databases. The Golden State Killer case is a famous example where distant relatives' DNA profiles led investigators to a suspect [46].

1.3.1 Databases and Global Cooperation

The establishment of national and international DNA databases, such as the CODIS in the United States [27, 40], or the European Union's Prüm Treaty, has drastically enhanced the ability to link crimes, identify repeat offenders, and solve cross-border cases. As of 2025, the U.S. national database contains over 20 million offender profiles and has aided in hundreds of thousands of investigations [28].

1.3.2 Key Techniques in Forensic Genetics

Among the techniques employed, short tandem repeat (STR) analysis remains the gold standard for human identification in forensic laboratories worldwide. The advent of next-generation sequencing (NGS) technologies has further expanded the scope of forensic genetics, enabling the analysis of complex mixtures, degraded samples, and even predictive assessments related to biogeographic ancestry or physical traits [32, 56, 57]. More regarding key techniques will be discussed in Chapter 2.

1.4 CHALLENGES AND ETHICAL CONSIDERATIONS

Despite its strengths, the use of genetic data in forensic contexts is not without challenges. Many of those challenges are highlighted elsewhere in this chapter and in Chapter 2. The interpretation of low-template or mixed DNA profiles can be complex and contentious. Moreover, the expanding use of FIGG and large-scale genetic databases raises significant ethical, legal, and privacy concerns, necessitating careful regulation and oversight [9, 46, 56, 58].

Despite its power, forensic genetics faces several challenges. Some of these challenges include the following:

- Interpretation of Complex Mixtures: Crime scene samples may contain DNA from multiple individuals, making interpretation difficult and sometimes subjective [34].
- Low-Template and Degraded DNA: Minute or compromised samples risk stochastic effects and allelic drop-out, potentially leading to misinterpretation or wrongful conviction [28, 34].
- Contamination and Quality Control: Strict laboratory standards, accreditation, and chain-of-custody procedures are crucial to maintain the integrity of results [28].
- Privacy and Consent: The use of genetic databases, especially in FIGG, has raised significant privacy and civil liberties concerns.

Several high-profile legal cases have questioned the limits of law enforcement's access to consumer DNA data [46].

- Ethical Use of Predictive Genetics: Emerging technologies can predict ancestry, appearance, and even some behavioral traits, raising profound ethical questions about surveillance, discrimination, and consent [46].

Chapter 2 will discuss in greater detail some of the legal and ethical issues faced due to legal challenges and incidents involving corruption that have resulted in wrongful convictions. Despite advances in DNA science and technology, vulnerabilities remain that underscore the importance of not relying *solely* on DNA evidence to prove or disprove an inquiry. Here, the author emphasizes the importance of thorough investigative work and not relying solely on forensic evidence to prove or disprove a case.

1.5 SUMMARY

Forensic genetics continues to transform the landscape of criminal justice, offering both remarkable opportunities and profound responsibilities. As methodologies grow ever more sophisticated, ongoing research, ethical vigilance, and public engagement will be crucial in ensuring that the power of the genome is harnessed responsibly in the service of truth and justice. Through this introduction and overview of key history, fundamentals, techniques, methods, and challenges related to forensic genetics in the criminal justice system, it should serve to enlighten the reader on the importance of keeping an open mind while conducting any forensic investigation and paying critical attention to detail to every aspect to preserve the integrity of each forensic investigation.

High-profile cases illustrate the power and pitfalls of forensic genetics and emerging technologies. Advancements in the evolving role of forensic genetics in the justice system represent an increasingly important responsibility to the public, ensuring the integrity of every investigation and examination entrusted.

REFERENCES

[1] G. M. Laporte, "Wrongful Convictions and DNA Exonerations: Understanding the Role of Forensic Science." *National Institute of Justice Journal*, 2017. [Online]. Available: https://nij.ojp.gov/topics/articles/wrongful-convictions-and-dna-exonerations-understanding-role-forensic-science

[2] U. S. D. o. Justice, "United States Department of Justice Interim Policy Forensic Genetic Genealogical DNA Analysis and Searching (FGGS)." *U.S. Department of Justice*. September 24, 2019. Available: https://www.justice.gov/olp/page/file/1204386/download (accessed).
[3] U. S. D. o. Justice, "Advancing Justice Through Dna Technology: Using dna to Solve Crimes." *U.S. Department of Justice*. Available: https://www.justice.gov/archives/ag/advancing-justice-through-dna-technology-using-dna-solve-crimes
[4] N. I. o. Justice, *What Every Law Enforcement Officer Should Know About DNA Evidence*. National Institute of Justice, 1999.
[5] D. Lithwick, "A Horrifying Miscarriage of Justice in North Carolina; How Many Times Was Justice Antonin Scalia Wrong about Henry Lee McCollum and the Death Penalty?." [Online]. Available: https://slate.com/news-and-politics/2014/09/henry-lee-mccollum-cleared-by-dna-evidence-in-north-carolina-after-spending-30-years-on-death-row.html
[6] M. Weigel, "DNA Evidence, Forensic Science and Confessions That Corrupt." *The Journalist Resource*, 2013. [Online]. Available: https://journalistsresource.org/criminal-justice/confessions-corrupt-evidence-cases/
[7] A. Mobley, "New DNA Testing in West Memphis 3 Case May Exonerate Convicted, Find 'Real Killer' in 2025." [Online]. Available: https://katv.com/news/local/new-dna-testing-in-west-memphis-3-case-may-exonerate-convicted-find-real-killer-damien-echols-jason-baldwin-jessie-misskelley-satanic-child-murder-ritual-1993-arkansas-m-vac-system-patrick-benca-dan-stidham-wm3-paradise-lost-harvest-of-innocence-project
[8] B. Paviour, "A Crime Lab Scientist's Work Exonerated 13 People." *But some say she altered evidence*. [Online]. Available: https://www.npr.org/2023/10/23/1208061548/a-crime-lab-scientists-work-exonerated-13-people-but-some-say-she-altered-eviden
[9] P. Chowdri, "Emerging Tech and Law Enforcement: Forensic Genetic Genealogy." [Online]. Available: https://www.police1.com/investigations/emerging-tech-and-law-enforcement-forensic-genetic-genealogy
[10] J. J. McShane, "Balancing Privacy and Public Safety: The Role of 23andMe's DNA Database in Criminal Investigations." [Online]. Available: https://thetruthaboutforensicscience.com/balancing-privacy-and-public-safety-the-role-of-23andmes-dna-database-in-criminal-investigations/

[11] Office of Cybersecurity Energy Security and Emergency Response. "Authorities and Roles." *U.S. Department of Energy*. Available: https://www.energy.gov/ceser/authorities-and-roles (accessed).

[12] N. H. Lents, "Use of Familial DNA in An Investigation Can Be Intrusive. But a middle Ground is Possible; New Jersey Police's Use of an Infant's Blood Sample to ID a Suspect May Point to Tremendous Potential for Solving Cold Cases, But it Has Rightfully Alarmed Privacy Activists." *Think*. [Online]. Available: https://www.nbcnews.com/think/opinion/new-jersey-polices-use-babys-dna-undermines-publics-trust-rcna43996

[13] E. Seiwell, "Lawyer for Wife of Accused Gilgo Beach Killer Rex Heuermann Says Netflix Series Raises 'Questions' about DNA." [Online]. Available: https://www.nydailynews.com/2025/04/15/lawyer-for-wife-of-accused-gilgo-beach-killer-rex-heuermann-says-netflix-series-raises-questions-about-dna/

[14] M. Marin, "Former Philly Homicide Detectives Acquitted of Perjury in Murder Investigation — But Jury Finds Two Lied about DNA Knowledge." [Online]. Available: https://www.msn.com/en-us/news/crime/former-philly-homicide-detectives-acquitted-of-perjury-in-murder-investigation-but-jury-finds-two-lied-about-dna-knowledge/ar-AA1BMPS4?ocid=BingNewsVerp

[15] S. Ceasar, "Defense Attempts to Discredit DNA Evidence in Grim Sleeper Trial." [Online]. Available: https://www.latimes.com/local/lanow/la-me-ln-grim-sleeper-dna-20160224-story.html

[16] S. Boyd, "Colorado Bureau of Investigation under Fire for Not Notifying Defendants Evidence Used Against Them May Have Been Falsified." [Online]. Available: https://www.cbsnews.com/colorado/news/colorado-bureau-investigation-defendants-evidence-used-against-may-have-been-falsified/

[17] A. Martinez, "Texas Death Row Inmate David Wood 'Desert Killer' Maintains Innocence Days Before Execution." [Online]. Available: https://theramswire.usatoday.com/story/news/crime/2025/03/06/david-leonard-wood-maintains-innocence-days-before-scheduled-execution/80524604007/

[18] A. L. Myers, "Texas Court Issues Rare Stay of Execution for Death Row Inmate Who Fought for DNA Testing." [Online]. Available: https://www.usatoday.com/story/news/nation/2025/03/11/david-leonard-wood-execution-texas/82286994007/

[19] J. Hall and J. Masterson. "Obstruction of Justice: Breaking Down 18 U.S.C. § 1503." *Law Offices of Masterson Hall*. Available: https://www.mastersonhall.com/obstruction-of-justice-breaking-down-18-usc-1503/#Evidence_Destruction_Methods (accessed).

[20] N. Elstor, "How Forensic DNA Evidence Can Lead to Wrongful Convictions." *Jstore Daily*, 2017. [Online]. Available: https://daily.jstor.org/forensic-dna-evidence-can-lead-wrongful-convictions/
[21] J. Jojola, "Conviction Overturned Due to DNA Evidence Scandal Could Be Only the First." April 15, 2025 [Online]. Available: https://www.9news.com/article/news/local/dna-evidence-conviction-overturned-michael-clark-missy-woods/73-0b7f8b31-0819-4204-be7c-6f436a174655
[22] D. o. S. Staff. "Causes of Wrongful Conviction." *Western Michigan University, Department of Sociology*. Available: https://wmich.edu/sociology/causes-wrongful-conviction
[23] E. Levenson, "3 men who have spent over two decades in prison appeal murder convictions on new DNA evidence." [Online]. Available: https://www.cnn.com/2023/08/15/us/pennsylvania-murder-appeal-dna/
[24] N. I. o. Justice, "Using DNA to Solve Cold Cases." *National Institute of Justive*, 2002. [Online]. Available: https://www.ojp.gov/pdffiles1/nij/194197.pdf
[25] G. Staff. "Open-Source DNA Databases: What They Are and How to Use Them." *Genomelink*. February 11, 2023. Available: https://genomelink.io/blog/open-source-dna-databases (accessed).
[26] D. Chacon-Cortes and and L. R. Griffiths, "Methods for Extracting Genomic DNA from Whole Blood Samples: Current Perspectives." *Journal of Biorepository Science for Applied Medicine*, vol. 2, no. null, pp. 1–9, 2014, doi: 10.2147/BSAM.S46573
[27] A. Kovacs, "Law Enforcement Access to DNA Databases: Privacy and Legal Implications." ed. The DNA Tests, 2025.
[28] J. Butler, *Forensic DNA Typing: Biology, Technology, and Genetics of STR Markers*, 2nd ed. Elsevier Academic Press, 2005.
[29] K. Lo. "Whose DNA is It Anyway? Legal Challenges that Arise from the Use of Genetic Genealogy in Criminal Investigations." *Richmond Journal of Law and Technology*. University of Richmond School of Law. April 26, 2024. Available: https://jolt.richmond.edu/2024/04/26/whose-dna-is-it-anyway-legal-challenges-that-arise-from-the-use-of-genetic-genealogy-in-criminal-investigations/ (accessed).
[30] L. Hamers, "When DNA Lies: The Hidden Flaws in Forensic Science." *SciTech Daily*, 2024. [Online]. Available: https://scitechdaily.com/when-dna-lies-the-hidden-flaws-in-forensic-science/
[31] P. A. Smith, "When DNA Implicates the Innocent." *Scientific American*, no. June 2016, 2016. [Online]. Available: https://www.scientificamerican.com/article/when-dna-implicates-the-innocent/
[32] D. J. R. J. W. M. Ruitberg, and C. H. Butler, "STRBase: A Short Tandem Repeat DNA Database for the Human Identity Community." *Nucleic Acids Research*, vol. 29, no. 1, pp. 320–322, 2001.

[33] P. Wagner, "'Shameful': D.C. Murder Case Dismissed After DNA Evidence Lost." [Online]. Available: https://www.nbcnewyork.com/news/national-international/dc-murder-case-dismissed-dna-evidence-lost/6290680/
[34] A. J. J. P. Gill, and D. J. Werrett, "Forensic Application of DNA 'Fingerprints'." *Nature*, vol. 318, no. 6046, pp. 577–579, 1985.
[35] J. McShane. "DNA Interpretation: The Prosecutor's Fallacy." [Online]. Available: https://thetruthaboutforensicscience.com/dna-interpretation/
[36] A. Ossola, "DNA Evidence Is Not Foolproof; Believing it blindly could put the wrong people in jail." *Popular Science*, 2015. [Online]. Available: https://www.popsci.com/dna-evidence-not-foolproof/#:~:text=Juries%20and%20lawyers%20alike%20generally%20consider%20DNA%20evidence,less%20reliable%20and%20objective%20than%20most%20people%20think
[37] D. Wu, "Crime Lab Analyst Altered DNA Evidence in Hundreds of Cases, Prosecutors Say." [Online]. Available: https://www.msn.com/en-us/news/crime/crime-lab-analyst-altered-dna-evidence-in-hundreds-of-cases-prosecutors-say/ar-AA1xLgMH?ocid=msedgntphdr&cvid=621d055d1812406e95ce1be7cf2dfc4a&ei=25
[38] J. J. McShane, "Contamination and Cross-Contamination: Understanding the Risks of DNA Evidence." [Online]. Available: https://thetruthaboutforensicscience.com/contamination-and-cross-contamination-understanding-the-risks-of-dna-evidence/
[39] K. Hall, "Concerned over DNA evidence, judge recommends new trial to Austin man on death row for 2009 slaying." *Austin American Statesman*, ed: Austin American Statesman, January 8, 2021. Available: https://www.statesman.com/story/news/2021/01/06/judge-in-travis-county-concerned-over-dna-evidence-grants-new-trial-austin-man-death-row/4128619001/
[40] W. R. Tuazon, R. Ansell, C.J. Guerrini, G.J. Zwenne, and B. Custers, "Law Enforcement Use of Genetic Genealogy Databases in Criminal Investigations: Nomenclature, Definition and Scope." *Forensic Science International: Synergy*, 2024, doi: 10.1016/j.fsisyn.2024.100460
[41] K. Miles, "The 27-Year-Old Cold Case That Will Put Familial DNA on Trial." [Online]. Available: https://downeast.com/issues-politics/dna-evidence/
[42] L. St. Angelo, "A 24-Year-Old Teacher Was Murdered in 1971. DNA On a Cigarette Butt Just Solved the Case." [Online]. Available: https://www.usatoday.com/story/news/nation/2023/02/22/cold-case-solved-dna-cigarette-rita-curran/11320406002/
[43] Staff. "New DNA Evidence Frees Hawaii Man After 30 Years in Prison for Murder." [Online]. Available: https://www.cbsnews.com/news/hawaii-man-gordon-cordeiro-freed-murder-30-years-dna-evidence/

[44] C. Nesi, "Inmate Gordon Cordeiro, free after 30 years thanks to new DNA evidence, struggles to adapt to modern times." [Online]. Available: https://nypost.com/2025/02/23/us-news/gordon-cordeiro-free-after-30-years-thanks-to-new-dna-evidence-struggles-to-adopt-to-modern-times/

[45] S. H. Hess Cordula, "Recovery of Trace DNA on Clothing: A Comparison of Mini-tape Lifting and Three Other Forensic Evidence Collection Techniques." *Journal of Forensic Sciences*, vol. 61, no. 1, 2017, doi: 10.1111/1556-4029.13246

[46] K. Kennett, "Forensic Genealogy: Some Serious Concerns." *Forensic Science International: Genetics*, vol. 36, pp. 203–209, 2018.

[47] P. Gleeson, "Bombshell Claims 'Botched' DNA Testing at Queensland Forensic Centre May Have Caused Hundreds of Flawed Prosecutions." [Online]. Available: https://www.skynews.com.au/australia-news/politics/bombshell-claims-botched-dna-testing-at-queensland-forensic-centre-may-have-caused-hundreds-of-flawed-prosecutions/news-story/20c19d86511a2abd1a62e378c47a046b

[48] M. University. "How Is DNA Profiling Used to Solve Crimes?" Maryville University. September 22, 2021. Available: https://www.cliffsnotes.com/study-notes/21242109#:~:text=9/22/21%2C%203,putting%20innocent%20people%20behind%20bars (accessed).

[49] L. Cler, D. Bu, C. Lewis, and D. Euhus, "A Comparison of Five Methods for Extracting DNA from Paucicellular Clinical Samples." *Molecular and Cellular Probes*, vol. 20, no. 3, pp. 191–196, 2006, doi: 10.1016/j.mcp.2005.12.003

[50] C. J. J. Huijsmans, J. Damen, J. C. van der Linden, P. H. M. Savelkoul, and M. H. A. Hermans, "Comparative Analysis of Four Methods to Extract DNA from Paraffin-embedded Tissues: Effect On Downstream Molecular Applications." *BMC Research Notes*, vol. 3, no. 1, p. 239, 2010, doi: 10.1186/1756-0500-3-239

[51] M. Andrews, "Murder Case of Heather Bogle Unraveled Errors and Corruption in Sandusky County." [Online]. Available: https://www.facebook.com/wtol11/posts/11-investigates-revisited-the-murder-of-heather-bogle-10-years-after-her-death-w/1125825892920600/

[52] K. Silva, "Corrupt Victoria Police Officer Jailed After Falsifying Evidence in Drug Squad Investigation." [Online]. Available: https://www.msn.com/en-au/news/australia/corrupt-victoria-police-officer-jailed-after-falsifying-evidence-in-drug-squad-investigation/ar-AA1ujX4n?ocid=BingNewsVerp

[53] K. Richmond, "AI-driven DNA Evidence Analysis Has a Transparency Problem." *The Conversation*, 2020. [Online]. Available: https://thenextweb.com/news/ai-driven-dna-evidence-analysis-has-a-transparency-problem

[54] M. E. Goldberg, "13-Year-Old Pulled Out Her Own Hair to Leave DNA Evidence at Home of Alleged Sexual Assaulter, Police Say." [Online]. Available: https://people.com/13-year-old-pulled-out-her-hair-to-leave-dna-evidence-at-home-of-alleged-assaulter-11714179
[55] D. A. Paprota, "The Influence of Higher Education on Law Enforcement Entry Level Examination Outcomes." D. Seton Hall University, New Jersey, 2012. [Online]. Available: https://trident.idm.oclc.org/login?url=https://www.proquest.com/dissertations-theses/influence-higher-education-on-law-enforcement/docview/1033649479/se-2?accountid=28844, https://YV9QF3BQ4D.search.serialssolutions.com?ctx_ver=Z39.88-2004&ctx_enc=info:ofi/enc:UTF-8&rfr_id=info:sid/ProQuest+Dissertations+%26+Theses+Global&rft_val_fmt=info:ofi/fmt:kev:mtx:dissertation&rft.genre=dissertations&rft.jtitle=&rft.atitle=&rft.au=Paprota%2C+David+A.&rft.aulast=Paprota&rft.aufirst=David&rft.date=2012-01-01&rft.volume=&rft.issue=&rft.spage=&rft.isbn=978-1-267-53637-2&rft.btitle=&rft.title=The+Influence+of+Higher+Education+on+Law+Enforcement+Entry+Level+Examination+Outcomes&rft.issn=&rft_id=info:doi/
[56] M. J. B. Catherine, A. Ball, J. Byrnes, P. Carbonetto, K. G. Chahine, R. E. Curtis, J. M. Granka, E. Han, E. L. Hong, A. R. Kermany, N. M. Myres, K. Noto, J. Qi, K. Rand, D. Barry Starr, Y. Wang, L. Willmore, and A. B. Wolf, "AncestryDNA Matching White Paper." *Ancestrycdn.com*, 2025. Available: https://www.ancestrycdn.com/support/us/2025/01/2025matchingwhitepaper.pdf
[57] A. Staff. "Ancestry Guide for Law Enforcement." *Ancestry.com*. Available: https://www.ancestry.com/c/legal/lawenforcement?ranMID=50138&ranEAID=3420415&ranSiteID=Gl6mUn9YQ68-5h6X0CS3C7JHNICa28bb5Q&o_xid=0003420415&o_lid=0003420415&o_sch=Affiliate%20External
[58] A. Lee. "Can Law Enforcement Use Your Ancestry DNA Results? | Genetic Genealogy." Available: https://www.youtube.com/watch?v=gPM7qTelfwY

CHAPTER 2

How DNA Profiling Is Used in the Criminal Justice System

Richard Allen Anderson

CONTENTS

2.1 INTRODUCTION

Oh, let us count the ways! As DNA discoveries and technologies have increased, both law enforcement and criminal defense advocates have increased the use of DNA collection and testing to identify, prosecute, and exonerate the accused in the criminal justice system. Law enforcement uses DNA in their efforts to identify and confirm suspects at crime scenes and, even more directly, to determine whether individuals have handled

 DOI: 10.4324/9781032714998-2

implements used in crimes. Defense law firms use DNA to defend the accused and seek to overturn convictions. Cold case investigators often examine decades-old evidence and, in some cases, use DNA to compare it with genealogical databases and identify suspects.

DNA is used to generate leads and thus narrow down a list of potential suspects and often directly identify perpetrators. Crimes DNA is used to help solve not only the most heinous murders and rapes but also lesser serious crimes such as burglaries, thefts, and other property crimes. The U.S. Department of Justice/FBI (Federal Bureau of Investigation) maintains a Combined DNA Index System (CODIS) of genetic profiles of persons arrested or convicted for felonies and some misdemeanors, and state and local criminal justice labs can submit DNA samples collected from crime scenes for analysis and comparison to DNA profiles in the CODIS to identify suspects [1].

This chapter will discuss DNA profiling, some of the history behind the use of DNA in the criminal justice system, legal issues, and even delve into some examples of issues and concerns when DNA is intentionally or inadvertently corrupted, or "planted" to stage a crime scene. Since DNA is much like a fingerprint, that it can, with reasonable accuracy, identify one individual [2, 3], it is not flawless by any means and mistakes can be made in the laboratory [4], and it can be used by corrupt individuals, both civilians and law enforcement [4–9], to make or break a case, and lead either to wrongful convictions of the accused. So, let's begin our journey into this chapter with neither the eyes of law enforcement, a prosecutor, or the defense, but *an open mind*, and let's explore what the use of DNA brings to the table.

2.2 HISTORY, LEGALITY, AND CONCERNS OF USING DNA IN THE CRIMINAL JUSTICE SYSTEM

Since the discovery of the many uses of DNA profiling within the criminal justice system, and its first use in 1989 [2], there have been many landmark achievements, such as the ability to, with a high degree of accuracy, identify suspects in horrendous murder and sexual assault cases, and even in low-impact property crimes [2, 10–14]. Hence, identifying perpetrators using DNA obtained from weapons and other instruments of the crime has led to convictions, often beyond a reasonable doubt, in many cases [2, 11, 12, 14]. But what happens when a person's DNA is at the scene of a crime, but in reality, the person may not have committed the crime? The answers to that question can become very complex, and this chapter cannot address all the variables or debates that can arise in such circumstances. However, some light can be shed on a variety of circumstances, in some instances, taken from real-world examples, that can lead to arguments both for the prosecution and exoneration of an individual accused of a crime [6, 7, 15–21].

2.2.1 DNA Profiling

What is DNA profiling? In brief, DNA profiling, which is also called genetic fingerprinting, is a forensic laboratory process used to determine a person's deoxyribonucleic acid characteristics, which are unique to that individual. Evidence obtained from "DNA profiling" or "DNA fingerprinting" is commonly used as evidence in criminal proceedings, along with other evidence, to determine the guilt or innocence of a person accused of a crime [12].

Depending on the DNA sample, there are various methods used to extract DNA. Some of these methods include mechanical, chemical, and enzymatic lysis, precipitation, purification, and concentration. The specific method used to extract the DNA, such as phenol-chloroform extraction, alcohol precipitation, or silica-based purification [22–24].

Some of the many different methods include lysis, precipitation, purification, and concentration. The lysis method entails breaking open the cells to release the DNA. Through the precipitation method, proteins and other contaminants must be removed from the released DNA. DNA will form solid matter at the bottom of a solution, while contaminants will remain in the liquid. Purification is performed after DNA is precipitated and refined through column-based methods. Finally, concentration results from the removal of any remaining liquids after a freeze-drying process [23, 24].

There are other DNA extraction processes not mentioned in this overview, but the end result of obtaining a unique DNA "fingerprint" is purportedly unique to one individual [2, 12, 22]. However, as this chapter discusses, there are several flaws and issues with DNA and its uses within the scientific and criminal justice communities. It is not the one-stop, all-inclusive crime-solving tool; instead, its techniques, uses, and those responsible for the integrity of evidence have rendered it increasingly challenged [3, 7, 18, 25–27].

2.2.2 History

Since the early uses of DNA in the criminal justice system in 1989 and throughout the 1990s [2], DNA use has evolved from the once-thought nail that sealed the coffin shut on criminal cases through legal challenges regarding its collection, use, and yes, even issues of corruption in criminal convictions [4, 5, 18, 25, 28–31]. However, as criminal defense practitioners expanded their understanding of DNA and explored ulterior possibilities how their clients' DNA could be at a crime scene, or other circumstances such as the lack of DNA evidence or other circumstances, DNA, or the lack of, or unidentified DNA at a crime scene, has led to the exoneration of the accused, sometimes sadly, after decades of incarceration [6, 18–20, 32–34].

Early uses of DNA that advanced through the 1990s and resulted in the solving of criminal cases that otherwise may have never been solved [2, 11, 12, 14]. The U.S. Department of Justice created and maintains a nationwide forensic DNA database known as the CODIS. Both the federal government and all 50 states plus the District of Columbia upload DNA profiles to CODIS, and there are around 19 million or more DNA profiles in CODIS alone [2, 12]. As states and the federal government increase contributions to the various DNA profile databases, some jurisdictions have legislated that DNA samples from arrestees and convicted criminals for less serious crimes be collected and contributed to the state and federal databases. This, of course, increases the likelihood that unidentified perpetrators of more serious crimes might be arrested for lesser serious crimes in the future, and result in identifying suspects in more serious unsolved cases [2, 10, 12].

As uses of DNA increased over the years outside the criminal justice system, such as for ancestry genealogy research through companies such as Ancestry, 23andMe, and other private sector businesses where the public voluntarily submitted personal DNA samples, worldwide, by the 100's of thousands to aid in the search for long-lost relatives and the mysteries of their family lineage [35–37]. As well, DNA profiling has been done by the medical community in the name of genetic research to predetermine one's susceptibility to disabling diseases and in search of cures [38]. Despite the seemingly harmless and beneficial uses of DNA by the public, little known to many was that unscrupulous law enforcement would seek to tap into these treasure troves of DNA profiling through medical records and genealogical databases by posing as someone they were not, submitting DNA evidence from a crime scene, in order to find out who the crime scene DNA belongs to. Many of these instances were done without warrants and presented numerous Fourth Amendment issues (e.g., Maryland v. King, 2013) [26, 28, 35–37, 39, 40]. Some of these and other examples will be addressed later in this chapter.

2.2.3 Legality

For the above reasons, and to protect the privacy of their clientele, some of the genealogy research companies, such as Ancestry.com, formed strict anti-cooperation policies with law enforcement [36, 37] while others cooperated with law enforcement without their customers' knowledge [27, 28, 35, 41, 42]. While it is unscrupulous, akin to deception, for law enforcement to pose as someone they are not to attempt to use a private company's DNA database to solve crimes, leading to Fourth Amendment unlawful search and seizure challenges [26, 28, 37, 39], court decisions have paved the way for law enforcement to obtain subpoenas and warrants for the DNA data [2, 3, 27, 40].

2.3 DNA EVIDENCE IN THE COURTS

2.3.1 Issues and Concerns

Just as DNA can appear to demonstrate that an individual was "at a crime scene" or "handled an implement of a crime" or "had physical contact with" a crime victim per se, the presence of DNA at a crime scene, on an instrument used in the commission of a crime, or on the body of a crime victim alone is not enough alone to convict an accused of a crime [6, 7, 15, 16, 32]. Some may argue the author's point, but let's start by looking at a rather simple, hypothetical crime:

A murder has been committed. In this case, whether the murder weapon was a gun, a knife, or a hatchet makes no difference. The body was discovered after several days of searching for the missing victim. Furthermore, the murder weapon was discovered after an anonymous tip, and it was matched to the crime. DNA evidence belonging to John Doe Smith was discovered on the murder weapon. Mr. Smith admittedly owned the murder weapon, whether it was a kitchen knife, the hatchet from his barn, or a gun he usually kept in his truck. To top it off, Mr. Smith had an argument and an ongoing dispute with the victim a day before the victim went missing, witnessed by several people, and his only alibi was that he was home alone during the approximate time of the murder with no witnesses to corroborate his alibi. In addition, the victim lived alone, outside the city, on several acres of land where their home was not within view of the public road with easy public access, where Mr. Smith could have gone to their home unobserved. On top of all that, the DNA on the murder weapon matched Mr. Smith, and no fingerprints were able to be obtained from the murder weapon, just smudges. It was suggested during the trial that Mr. Smith must have worn gloves while committing the crime when the absence of fingerprints on the weapon was deliberated.'

In the above hypothetical case, all evidence points to Mr. Smith. He had the motive, means, and opportunity to commit murder; the weapon used belonged to Mr. Smith, and Mr. Smith's DNA was on the weapon. An open and shut case? Ponder this case for a few moments with your law enforcement hat on, then ponder it for a few moments with your defense attorney's hat on. What arguments can you come up with for both sides? Now consider that the victim had made other enemies, not just Mr. Smith.

One of those enemies wanted to kill the victim and was an acquaintance of Mr. Smith, and knew Mr. Smith well enough, and knew where Mr. Smith kept the tool eventually used in the crime (gun, knife, hatchet, etc.). While many of the readers may have obtained the conviction of Mr. Smith at a jury trial no less, and Mr. Smith is rotting away in a prison cell, maybe even on death row, the real murderer was an intelligent, cunning person and wore gloves while handling the murder weapon. Wait a minute, right?! Why would Mr. Smith wears gloves while using a murder weapon that could easily be traced back to him?

As one can observe from this rather simple murder scenario, several issues can arise, leading to questions from both sides of the debate. In that scenario, Mr. Smith was actually set up for the crime, convicted, and sent to prison. Accompanying DNA evidence, there existed motive, means, and opportunity, the lack of a verifiable alibi, and other circumstantial evidence. But in this hypothetical case, an undiscovered reality existed that another person orchestrated and committed the crime, not Mr. Smith. Do these things really happen? Are innocent people sent to prison, possibly even executed, for crimes they did not commit? Sadly and soberingly, the answer is an astounding YES!

Despite the valuable tool DNA has become within the criminal justice system, the presence or absence of DNA should be sternly evaluated along with other factors for several reasons. The presence or absence of DNA alone should not be considered the sole nail in the coffin of guilt, but due diligence should be given to all physical, circumstantial, and mitigating circumstances in every case [3, 7, 43]. Yes, DNA is physical evidence, but both prosecutors and defense attorneys should consider all options.

Prosecutors can too often be prideful and narrowly focus on their conviction numbers and can become tainted into a mindset that everyone is guilty, and even ignore or cover up knowledge of extenuating circumstances, statements, or evidence they do not want the prosecution to have that may pose a reasonable doubt as to the guilt of an individual by a judge or jury. Yes, the author has directly observed this corruption in his career, and after voicing concerns, he resigned from an agency as a result. This author asserts that such actions by a prosecutor (and officers, deputies, or agents) *are indeed* corrupt and unlawful, and if exposed and adjudged by a higher authority, criminal charges could ensue. At the very least, this sort of behavior is dishonest. Prosecutors who do this *do not* just do this in *one* case; actions such as this are *habitual* and affect many cases! An even bigger systematic problem exists when higher authorities refuse to investigate agencies and prosecutors when presented with an allegation and evidence.

Defense attorneys too often focus on making plea deals and getting the best "deal" they can, coupled with an admission of guilt for their clients. Yes, possibly the majority of the time, their clients are indeed guilty. However, this author again asserts that defense attorneys should arm themselves with advanced knowledge, understanding, and the development of increased skills beyond those learned in law school to present higher degrees of intellect and skill through continued training and case studies in forensic science. Sometimes guilty people get away with their crimes, but what is far worse is when the criminal justice system fails an innocent person by sending that innocent person to prison, or worse yet, to execution! More and more, DNA is being used in older cases, leading to the exoneration of many people unjustly incarcerated, often for several decades [4, 7, 19, 44–47].

The following are some real-world examples of cases corrupted by law enforcement and accusers using DNA (planting DNA evidence), and some that were later exonerated through the use of DNA evidence:

Case # 1: "The convictions of two mentally disabled half-brothers were vacated and the two men were ordered released by Superior Court Judge Douglas Sasser in North Carolina on Tuesday. They were freed from prison Wednesday. Henry Lee McCollum, 50, had been on death row for 30 years, longer than anyone else in North Carolina history. He and Leon Brown, 46, who was serving a life sentence, were convicted for the 1983 rape and murder of an 11-year-old girl. DNA evidence implicated another man, a known sex offender whom the police had not investigated, despite the fact that he lived next to the crime scene" [46].

Case # 2: In this decades-old murder case, Doug and Peggy Ryen, their 10-year-old daughter Jessica, and neighbor Chris Hughes (11) were stabbed to death, while 8-year-old son Josh Ryen managed to survive. A man who now appears to be innocent has been on death row for over 40 years for these crimes, despite mounting evidence suggesting he was not involved. DNA testing on evidence from the crime scene, including a tan T-shirt and hairs found in the victims' hands, did not match the convicted man Cooper, and no African-American DNA was found at the scene. Most significantly, an orange towel taken by the murderers yielded a full DNA profile that belongs to neither Cooper nor any of the victims, but remains unmatched.

Meanwhile, a different suspect, a convicted murderer and white man—who had been released less than a year before the killings—was implicated by his girlfriend, who reported he returned home bloody on the night of the murders. The sheriff's office discarded this different suspect's bloody coveralls, without so much as testing them, after arresting Cooper, not wanting complications or counternarratives to their case. Recent witness testimony has emerged with people willing to testify that this other suspect confessed to the murders, with one saying he bragged about "butchering" an entire family. While this second suspect's DNA also doesn't match the towel evidence, and he denies involvement, former FBI officials believe the case for Cooper's innocence continues to strengthen [19].

Case # 3: March 24, 2025, "The U.S. Supreme Court denied review of Areli Escobar's most recent appeal of his murder conviction, which argued for relief based on the fact that the Texas prosecutor had confessed error over the misleading use of inconclusive DNA evidence at trial. In his petition, Mr. Escobar's legal team said Travis County prosecutors had relied heavily at trial on compromised evidence analyzed by the Austin Police Department's crime laboratory. Additional DNA testing undertaken just days after Mr. Escobar's 2011 trial and hidden from the defense until 2017 revealed that the DNA testing presented at his trial was ultimately inconclusive,a fact that Mr. Escobar's jury never heard" [48].

Case # 4: "A police officer was jailed for three years after faking DNA and handwriting test results while working in the drug squad. County Court judge Michael Cahill labelled the actions of Jye Symes an abuse of power, and said it could have resulted in a woman wrongly being sent to jail. In his judgement Judge Cahill stated,'Corrupt police misconduct, as yours was, has significant social consequences. It undermines public confidence in the police force.' Symes had been working in Victoria Police's drug squad during an investigation into a drug safehouse in Cairnlea, in Melbourne's west" [5].

Case # 5: According to an affidavit filed in Jefferson County Court, Yvonne "Missy" Woods, 64, who worked for the Colorado Bureau of Investigation from 1994 to 2023, omitted signs of contamination introduced during the DNA testing process and incorrectly reported that some samples did not contain male DNA in analyses performed for cases including murder and sexual assault. Questioned by investigators, Woods allegedly said she falsified results to move cases along quickly and avoid additional work.

Woods, who retired in 2023 after the bureau began investigating her work, tampered with test data in over 1,000 cases, the Colorado Bureau of Investigation announced in August of 2025. The revelation rattled Colorado's courts, where attorneys are still grappling with the immense demands of retesting DNA samples and scrutinizing old cases... Woods is facing more than 100 felony charges. Before allegations of misconduct surfaced, Woods was a respected leader in the Colorado Bureau of Investigation lab, known as a fast worker who spent weekends on the job and regularly maxed out her overtime, the Denver Post reported. The Colorado Bureau of Investigation began investigating Woods in 2023 after an intern there working on a research project discovered missing data in test records entered by Woods, according to the affidavit [8].

What do you, the reader, observe as a common denominator in each case example presented above? Might the author suggest the common denominator is that of dishonesty, of corruption, and the author is certain the reader can form their own opinions. What lessons can one take away from these, and countless other real-world examples not presented in this chapter, but easily discoverable through a common Google or law library search? Might the author again suggest that some of these lessons may include that not all law enforcement operates with the integrity expected by the public, and that officers, investigators, and prosecutors can, and in many cases, are just as corrupt as any other perpetrator of a crime? Any time human beings are involved in anything, the potential for corruption is high just by the very nature of humans to lie, cheat, steal, even murder, to obtain coveted notoriety, promotions, close cases, or for simple bragging rights. These considerations present dangerous precedents within the criminal justice system, and from the author's observations throughout his career, he would suggest watch out for the officers that are set on getting x number of DUI's this weekend as they stalk the local bar around closing time, or the prosecutor that buries evidence, statements, or other facts surrounding a case that may be used by the defense. Get rid of them! I guarantee those exist in every city, county, department, and the majority of prosecutor's offices around the country!

It is the responsibility of EVERY Officer, Deputy, and Agent in law enforcement, and EVERY Attorney engaged as an Officer of the Court in the criminal justice system, to personally uphold the integrity expected of them by the public, and to act as watchdogs and whistleblowers to root out all forms of corruption within the criminal justice system. It is only by the sacrifices of honest persons who observe wrongdoings and corruptions, and who expose such corruption with bulldog tenacity, that the criminal justice system can hopefully, one day, be restored as a respected institution within our nation!

2.3.2 Addressing Statutes of Limitations Challenges

Legal issues, including statutes of limitations on certain crimes, present challenges. DNA obtained from the crime scene may provide a profile of the perpetrator; however, crimes less than murder often have just a few years in which charges can be filed against suspects. Statutes of limitations are in place for many reasons; just some of those include the effects of time regarding the accuracy of eyewitness testimony, and others [7, 49].

Statutes of limitations present serious issues regarding crimes where a suspect has not been identified. One solution that many law enforcement agencies consider is filing a "John Doe Warrant" for the person matching the DNA profile of the suspect, obtained during the investigation. By filing a "John Doe Warrant," agencies are able to meet the requirements of statutes of limitations prior to their expiration. State laws vary, and some states may address the use of "John Doe Warrants," while others may not [2].

2.4 DNA COLLECTION AND USE BY LAW ENFORCEMENT

This section summarizes some of the various DNA collection techniques used by law enforcement. DNA can be collected through a variety of means; here, though, are a few of the most common.

2.4.1 DNA Evidence Collection, Uses, and Techniques

Law enforcement typically collects DNA from suspected offenders, even before conviction, of crimes in their jurisdictions. DNA can be collected from persons or physical items for processing as evidence or to be added to DNA profiling databases for use within the criminal justice system or the private sector. For example, the criminal justice system may use DNA to prosecute or exonerate a suspect in a crime, or determine a possible suspect in an unsolved crime, or a cold case [2, 6, 7, 15–17, 32, 40]. Typically, when law enforcement overtly collects DNA from a suspect, arrestee, or person convicted of a crime, it is standard to obtain a buccal swab of the cheek to submit to the crime lab for analysis and cataloging into the state or local DNA database, or contribute to a federal database [3, 43].

Crime scene investigators can collect and preserve potential DNA evidence directly from a crime scene utilizing various techniques, depending on the type and materials of items suspected of having DNA, and preserving items for analysis by a crime lab. Typically, crime scenes can provide a plethora of potential for DNA collection based on the environment, conditions, human traffic within the environment, and evidence left behind within, on, or around the victim and scene.

Here are a few examples of how and where DNA in or around a crime scene can be collected by law enforcement. In instances where the crime victim is a person who had physical contact with the perpetrator, skin cells, bodily fluids, and hair left behind by the suspect are excellent sources of DNA, and proper care and preservation techniques should be taken to collect and preserve for the crime lab any trace of DNA left behind [50]. Various methods can be used in the crime scene, and in a crime lab, such as taking swabs for DNA samples from items believed handled by the suspect, collecting sheets, clothing, and other materials from the scene, and preserving them for lab analysis [22, 27, 50].

When a forensic investigator collects DNA samples from a crime scene, an additional process of collecting DNA samples from other people who have had access to the areas a perpetrator has accessed. An example could be a money box or safe that was accessed in the commission of a crime. Samples of DNA should be taken from persons who had legal, authorized access to the restricted area that the perpetrator accessed to eliminate those DNA profiles from the group of multiple DNA profiles that would be collected during the swab of the restricted area. In such situations, those who have had legal access can be separated from DNA profiles that may be unidentified. The unidentified DNA sample (in cases that it is not an inside job) can be compared to persons of interest or local and federal DNA databases [10, 50].

In some instances, a person of interest might not be cooperative; however, a relative of the person of interest might be willing to submit their DNA to aid law enforcement in determining whether their relative is or is not a viable person of interest and should be either investigated further or eliminated from a list of persons of interest [1, 27, 28, 35, 41, 42]. Various state or federal laws may apply regarding the collection and use of this or other techniques and uses of DNA in criminal cases, and all laws and court decisions (case laws) must be adhered to, in order to protect the integrity of any investigation [26, 28, 37].

2.4.2 Quick DNA Test Kits

The value of using DNA in the criminal justice system cannot be overstated.

When law enforcement collects, preserves, and profiles DNA from a crime scene, yet has no known suspects, the DNA profile of the unknown suspect is maintained in a federal and state DNA databases until either new information is received or the person it is matched to is arrested on another unrelated crime and their DNA is matched to the unknown suspect's DNA in the system. Many crimes have been solved, including cold cases, through this method [26].

Cold case investigations have turned to using DNA obtained from the crime scene to compare with DNA obtained through the arrest and collection of a DNA sample from a suspect in an entirely different crime. Furthermore, investigators have turned to genealogical DNA collected databases and comparison of DNA from relatives of persons of interest to narrow the search for suspects in cold cases or resolve old crimes altogether. As we have lightly addressed thus far, and will address in greater detail to follow, there are issues, concerns, and legal challenges in the use of biogeographic DNA databases such as Ancestry.com, 23andme.com, and others.

2.4.3 DNA in the Crime Lab

Within the crime laboratory setting, the analysis of DNA evidence follows rigorous scientific protocols to ensure accuracy, reproducibility, and the integrity of results. Upon receipt of DNA samples—whether from buccal swabs, biological fluids, hair, skin cells, or other trace evidence—laboratory personnel adhere to strict chain-of-custody procedures and apply validated methodologies for DNA extraction, quantification, amplification, and profile interpretation.

Modern crime labs utilize sophisticated techniques such as Polymerase Chain Reaction (PCR), Short Tandem Repeat (STR) analysis, and, in some cases, mitochondrial DNA (mtDNA) sequencing to generate unique DNA profiles. These profiles are then compared to reference samples or searched against national databases like the CODIS to identify suspects, link crimes, or exclude individuals from suspicion. Quality assurance measures, such as the use of positive and negative controls, proficiency testing of analysts, and adherence to accreditation standards (e.g., ISO/IEC 17025), are integral to maintaining the credibility and admissibility of DNA evidence in court.

Furthermore, laboratories are increasingly adopting rapid DNA technologies, which can expedite the turnaround time for results, allowing for faster investigative leads. However, every stage of the DNA analysis process—from evidence intake to reporting—must comply with state and federal laws, privacy protections, and best practices established by scientific and forensic organizations.

2.5 SUMMARY OF DNA EVIDENCE PRACTICES IN MODERN CRIME LABS

An overview of current techniques, quality measures, and legal considerations is as follows: Modern crime laboratories employ advanced DNA analysis techniques—including PCR, STR analysis, and mitochondrial DNA (mtDNA) sequencing—to create unique genetic profiles for forensic investigations. These profiles are compared with known samples or run through databases like CODIS to help identify suspects, connect crimes, or exonerate individuals.

The integrity and admissibility of DNA evidence rely on stringent quality assurance procedures such as using control samples, regular proficiency testing for analysts, and strict adherence to accreditation standards like ISO/IEC 17025. The adoption of rapid DNA technologies is further enhancing the speed of analyses, improving investigative timelines.

Every stage of DNA evidence processing must comply with legal requirements, privacy regulations, and established best practices from scientific and forensic authorities. The chapter emphasizes the critical role of continual oversight, robust protocols, and thorough documentation in upholding the reliability and credibility of DNA evidence in the criminal justice system.

The content provided offers a broad overview of DNA profiling, its history, and legal considerations in criminal justice. Implementing the above suggestions would enhance clarity, precision, and readability while maintaining the section's substance and formatting.

REFERENCES

[1] J. W. Hazel and E. W. Clayton. "Law Enforcement and Genetic Data." Available: https://www.thehastingscenter.org/briefingbook/law-enforcement-and-genetic-data/

[2] N. I. o. Justice, "Using DNA to Solve Cold Cases." *National Institute of Justice*, 2002. [Online]. Available: https://www.ojp.gov/pdffiles1/nij/194197.pdf

[3] N. I. o. Justice, *What Every Law Enforcement Officer Should Know About DNA Evidence*, National Institute of Justice, 1999.

[4] M. Weigel, "DNA Evidence, Forensic Science and Confessions that Corrupt." *The Journalist Resource*, 2013. [Online]. Available: https://journalistsresource.org/criminal-justice/confessions-corrupt-evidence-cases/

[5] K. Silva, "Corrupt Victoria Police Officer Jailed after Falsifying Evidence in Drug Squad Investigation." [Online]. Available: https://www.msn.com/en-au/news/australia/corrupt-victoria-police-officer-jailed-after-falsifying-evidence-in-drug-squad-investigation/ar-AA1ujX4n?ocid=BingNewsVerp

[6] G. M. Laporte, "Wrongful Convictions and DNA Exonerations: Understanding the Role of Forensic Science." *National Institute of Justice Journal*, 2017. [Online]. Available: https://nij.ojp.gov/topics/articles/wrongful-convictions-and-dna-exonerations-understanding-role-forensic-science
[7] N. Elstor, "How Forensic DNA Evidence Can Lead to Wrongful Convictions." *Jstore Daily*, 2017. [Online]. Available: https://daily.jstor.org/forensic-dna-evidence-can-lead-wrongful-convictions/
[8] D. Wu, "Crime Lab Analyst Altered DNA Evidence in Hundreds of Cases, Prosecutors Say." [Online]. Available: https://www.msn.com/en-us/news/crime/crime-lab-analyst-altered-dna-evidence-in-hundreds-of-cases-prosecutors-say/ar-AA1xLgMH?ocid=msedgntphdr&cvid=621d055d1812406e95ce1be7cf2dfc4a&ei=25
[9] M. E. Goldberg, "13-Year-Old Pulled Out Her Own Hair to Leave DNA Evidence at Home of Alleged Sexual Assaulter, Police Say." [Online]. Available: https://people.com/13-year-old-pulled-out-her-hair-to-leave-dna-evidence-at-home-of-alleged-assaulter-11714179
[10] U. S. D. o. Justice. "Advancing Justice Through DNA Technology: Using DNA to Solve Crimes." *U.S. Department of Justice*. Available: https://www.justice.gov/archives/ag/advancing-justice-through-dna-technology-using-dna-solve-crimes
[11] A. Rose, "A 1956 Double Murder Has Been Solved Using DNA Evidence. It May Be the Oldest Case Cracked Like This So Far, Investigator Says." [Online]. Available: https://www.cnn.com/2021/06/10/us/montana-1956-double-murder-solved-dna/
[12] M. University. "How Is DNA Profiling Used to Solve Crimes?" Maryville University. https://online.maryville.edu/blog/how-is-dna-profiling-used-to-solve-crimes/
[13] J. R. Miller, "DNA 'Enhancements' May Solve JonBenet Ramsey Killing 25 Years Later." [Online]. Available: Joshua Rhett Miller
[14] L. St. Angelo, "A 24-year-old Teacher Was Murdered in 1971. DNA On a Cigarette Butt Just Solved the Case." [Online]. Available: https://www.usatoday.com/story/news/nation/2023/02/22/cold-case-solved-dna-cigarette-rita-curran/11320406002/
[15] J. Jojola, "Conviction Overturned Due to DNA Evidence Scandal Could Be Only the First." [Online]. Available: https://www.msn.com/en-us/news/crime/conviction-overturned-due-to-dna-evidence-scandal-could-be-only-the-first/ar-AA1CZveV
[16] D. o. S. Staff. *Causes of Wrongful Conviction*. Western Michigan University, Department of Sociology. https://wmich.edu/sociology/causes-wrongful-conviction
[17] E. Levenson, "3 Men Who Have Spent Over Two Decades in Prison Appeal Murder Convictions On New DNA Evidence." [Online]. Available: https://www.cnn.com/2023/08/15/us/pennsylvania-murder-appeal-dna/

[18] P. A. Smith, "When DNA Implicates the Innocent." *Scientific American*, no. June 2016, 2016. [Online]. Available: https://www.scientificamerican.com/article/when-dna-implicates-the-innocent/
[19] N. Kristof, "Is an Innocent Man Still Languishing on Death Row? We've Seen the DNA Evidence in the Kevin Cooper Case. It Points Elsewhere." [Online]. Available: https://www.nytimes.com/2021/01/23/opinion/sunday/kevin-cooper-dna.html
[20] A. L. Myers, "'I'm Angry': Death Row inmate David Leonard Wood Says He's Innocent as Execution Nears." [Online]. Available: https://bearswire.usatoday.com/story/news/2025/03/05/death-row-inmate-david-leonard-wood-is-angry-over-execution-interview/81651738007/
[21] A. Ossola, "DNA Evidence Is Not Foolproof; Believing it Blindly Could Put the Wrong People in Jail." *Popular Science*, 2015. [Online]. Available: https://www.popsci.com/dna-evidence-not-foolproof/#:~:text=Juries%20and%20lawyers%20alike%20generally%20consider%20DNA%20evidence,less%20reliable%20and%20objective%20than%20most%20people%20think
[22] D. Chacon-Cortes and L. R. Griffiths, "Methods for Extracting Genomic DNA from Whole Blood Samples: Current Perspectives." *Journal of Biorepository Science for Applied Medicine*, vol. 2, no. null, pp. 1–9, 2014, doi: 10.2147/BSAM.S46573
[23] L. Cler, D. Bu, C. Lewis, and D. Euhus, "A Comparison of Five Methods for Extracting DNA from Paucicellular Clinical Samples." *Molecular and Cellular Probes*, vol. 20, no. 3, pp. 191–196, 2006, doi: 10.1016/j.mcp.2005.12.003
[24] C. J. J. Huijsmans, J. Damen, J. C. van der Linden, P. H. M. Savelkoul, and M. H.A. Hermans, "Comparative Analysis of Four Methods to Extract DNA from Paraffin-embedded Tissues: Effect on Downstream Molecular Applications." *BMC Research Notes*, vol. 3, no. 1, p. 239, 2010, doi: 10.1186/1756-0500-3-239
[25] L. Hamers, "When DNA Lies: The Hidden Flaws in Forensic Science." *SciTech Daily*, 2024. [Online]. Available: https://scitechdaily.com/when-dna-lies-the-hidden-flaws-in-forensic-science/
[26] A. Kovacs, "Law Enforcement Access to DNA Databases: Privacy and Legal Implications." ed. The DNA Tests, 2025.
[27] U. S. D. o. Justice. "United States Department of Justice Interim Policy Forensic Genetic Genealogical DNA Analysis and Searching." *U.S. Department of Justice*. Available: https://www.justice.gov/olp/page/file/1204386/dl
[28] K. Lo. "Whose DNA is It Anyway? Legal Challenges that Arise from the Use of Genetic Genealogy in Criminal Investigations." Available: https://jolt.richmond.edu/2024/04/26/whose-dna-is-it-anyway-legal-challenges-that-arise-from-the-use-of-genetic-genealogy-in-criminal-investigations/

[29] P. Wagner, "'Shameful': D.C. Murder Case Dismissed after DNA Evidence Lost." [Online]. Available: https://www.nbcnewyork.com/news/national-international/dc-murder-case-dismissed-dna-evidence-lost/6290680/
[30] J. Kaiser, "New Federal Rules Limit Police Searches of Family Tree DNA Databases." *Science Insider*, 2019. [Online]. Available: https://www.science.org/content/article/new-federal-rules-limit-police-searches-family-tree-dna-databases
[31] M. Andrews, "Murder Case of Heather Bogle Unraveled Errors and Corruption in Sandusky County." [Online]. Available: https://www.wtol.com/article/news/investigations/11-investigates/heather-bogle-murder-case-unraveled-errors-and-corruption-sandusky-county-11-investigates/512-81fc746e-d4da-41cb-8ed9-9bd36f779d5d
[32] A. Mobley, "New DNA Testing in West Memphis 3 Case May Exonerate Convicted, Find 'Real Killer' in 2025." [Online]. Available: https://katv.com/news/local/new-dna-testing-in-west-memphis-3-case-may-exonerate-convicted-find-real-killer-damien-echols-jason-baldwin-jessie-misskelley-satanic-child-murder-ritual-1993-arkansas-m-vac-system-patrick-benca-dan-stidham-wm3-paradise-lost-harvest-of-innocence-project
[33] B. Paviour, "A Crime Lab Scientist's Work Exonerated 13 People. But Some Say She Altered Evidence." [Online]. Available: https://www.npr.org/2023/10/23/1208061548/a-crime-lab-scientists-work-exonerated-13-people-but-some-say-she-altered-eviden
[34] A. Martinez, "Texas Death Row inmate David Wood 'Desert Killer' Maintains Innocence Days Before Execution." [Online]. Available: https://theramswire.usatoday.com/story/news/crime/2025/03/06/david-leonard-wood-maintains-innocence-days-before-scheduled-execution/80524604007/
[35] A. Lee. "Can Law Enforcement Use Your Ancestry DNA Results? | Genetic Genealogy." Available: https://www.youtube.com/watch?v=gPM7qTelfwY
[36] M. J. B. Catherine, A. Ball, J. Byrnes, P. Carbonetto, K. G. Chahine, R. E. Curtis, J. M. Granka, E. Han, E. L. Hong, A. R. Kermany, N. M. Myres, K. Noto, J. Qi, K. Rand, D. Barry Starr, Y. Wang, L Willmore, and A. B. Wolf, "AncestryDNA Matching White Paper." *Ancestrycdn.com*, 2025. [Online]. Available: https://www.ancestrycdn.com/support/us/2025/01/2025matchingwhitepaper.pdf
[37] A. Staff, "Ancestry Guide for Law Enforcement." *Ancestry.com*. https://www.ancestry.com/c/legal/lawenforcement?ranMID=50138&ranEAID=3420415&ranSiteID=Gl6mUn9YQ68-5h6X0CS3C7JHNICa28bb5Q&o_xid=0003420415&o_lid=0003420415&o_sch=Affiliate%20External
[38] G. Staff, "Open-Source DNA Databases: What They Are and How to Use Them." *Genomelink*. Available: https://genomelink.io/blog/open-source-dna-databases

[39] K. Miles, "The 27-Year-Old Cold Case That Will Put Familial DNA on Trial." [Online]. Available: https://downeast.com/issues-politics/dna-evidence/
[40] N. H. Lents, "Use of Familial DNA in An Investigation Can Be Intrusive. But a Middle Ground is Possible; New Jersey Police's Use of An Infant's Blood Sample to ID a Suspect May Point to Tremendous Potential for Solving Cold Cases, But It Has Rightfully Alarmed Privacy Activists." *Think*. [Online]. Available: https://www.nbcnews.com/think/opinion/new-jersey-polices-use-babys-dna-undermines-publics-trust-rcna43996
[41] O. M. Tuazon, R. A. Wickenheiser, R. Ansell, C. J. Guerrini, G.J. Zwenne, and B. Custers, "Law Enforcement Use of Genetic Genealogy Databases in Criminal Investigations: Nomenclature, Definition and Scope." *Forensic Science International: Synergy*, 02/08/2024 2024, doi: 10.1016/j.fsisyn.2024.100460
[42] P. Chowdri, "Emerging Tech and Law Enforcement: Forensic Genetic Genealogy." [Online]. Available: https://www.police1.com/investigations/emerging-tech-and-law-enforcement-forensic-genetic-genealogy
[43] Property & Evidence Control, L. E. P. Center, Law Enforcement Policy Center, 2021. [Online]. Available: http://www.chrome-extension://efaidnbmnnnibpcajpcglclefindmkaj/ https://www.theiacp.org/sites/default/files/2021-03/Evidence%20Control%20Formatted%2003.03.2021.pdf
[44] Staff, "New DNA Evidence Frees Hawaii Man after 30 Years in Prison for Murder." [Online]. Available: https://www.cbsnews.com/news/hawaii-man-gordon-cordeiro-freed-murder-30-years-dna-evidence/
[45] C. Nesi, "Inmate Gordon Cordeiro, Free after 30 Years Thanks to New DNA Evidence, Struggles to Adapt to Modern Times." [Online]. Available: https://nypost.com/2025/02/23/us-news/gordon-cordeiro-free-after-30-years-thanks-to-new-dna-evidence-struggles-to-adopt-to-modern-times/
[46] D. Lithwick, "A Horrifying Miscarriage of Justice in North Carolina; How Many Times Was Justice Antonin Scalia Wrong about Henry Lee McCollum and the Death Penalty?" [Online]. Available: https://slate.com/news-and-politics/2014/09/henry-lee-mccollum-cleared-by-dna-evidence-in-north-carolina-after-spending-30-years-on-death-row.html
[47] J. McShane, "DNA Interpretation: The Prosecutor's Fallacy." [Online]. Available: https://thetruthaboutforensicscience.com/dna-interpretation/

[48] H. Bedard, "U.S. Supreme Court Denies Review in Texas Faulty DNA Evidence Case, Despite Prosecutor Confession of Error, Death Penalty Center." Accessed October16, 2025. Available: https://deathpenaltyinfo.org/u-s-supreme-court-denies-review-in-texas-faulty-dna-evidence-case-despite-prosecutor-confession-of-error
[49] P. Gleeson, "Bombshell Claims 'Botched' DNA Testing at Queensland Forensic Centre May Have Caused Hundreds of Flawed Prosecutions." [Online]. Available: https://www.skynews.com.au/australia-news/politics/bombshell-claims-botched-dna-testing-at-queensland-forensic-centre-may-have-caused-hundreds-of-flawed-prosecutions/news-story/20c19d86511a2abd1a62e378c47a046b
[50] S. H. Hess Cordula, "Recovery of Trace DNA on Clothing: A Comparison of Mini-tape Lifting and Three Other Forensic Evidence Collection Techniques." *Journal of Forensic Sciences*, vol. 61, no. 1, 2017, doi: 10.1111/1556-4029.13246

CHAPTER 3

Biogeographic Ancestry Testing

Richard Allen Anderson

CONTENTS

DOI: 10.4324/9781032714998-3

3.1 INTRODUCTION

Biogeographic ancestry testing (BAT), a cutting-edge field at the intersection of genetics and forensics, has emerged as a powerful tool for criminal investigators. This chapter provides a comprehensive review of BAT in the context of criminal justice, examining its methodologies, applications, limitations, and ethical considerations. The chapter aims to clarify how this technology aids in solving complex cases, identifying unknown individuals, and narrowing suspect pools [1–4].

3.2 HISTORY OF ANCESTRY ANALYSIS IN FORENSIC SCIENCE

The application of ancestry analysis in forensics dates back to the 1990s, with the use of mitochondrial DNA (mtDNA) to infer maternal lineage. Advancements in technology have since expanded the use of DNA beyond criminal investigations to consumer goods and various other fields [5].

3.2.1 Definition and Significance of BAT

BAT involves inferring an individual's ancestral origins by analyzing specific genetic markers associated with particular populations or geographic regions. This process plays a vital role in identifying human remains, reconstructing crime scenes, and providing investigative leads by offering insights into a person's origins [1, 4].

3.2.2 Methodologies and Technologies

BAT utilizes various methods and technologies, including:

- **Single-Nucleotide Polymorphisms (SNPs)**: These are variations in a single DNA base pair, offering insights into ancestry and population history [1, 3].

- **Microhaplotypes:** Combinations of SNPs located close together on a chromosome are proving increasingly valuable for fine-scale ancestry inference [3, 6].
- **Uniparental Markers (Y-DNA and mtDNA):** These markers, passed down through the male (Y-DNA) and female (mtDNA) lines respectively, provide insights into ancestry along a specific parental line [3, 7, 8].

3.2.3 Applications in Criminal Justice

BAT finds diverse applications in criminal justice, such as:

- **Identifying Unknown Remains:** Ancestry information helps narrow down potential identities of unidentified human remains, assisting in their identification [9].
- **Developing Investigative Leads:** Information about a suspect's ancestry can provide leads for investigators, directing their search and focusing their efforts [10–12].
- **Narrowing Suspect Pools:** When used in conjunction with other forensic techniques, BAT can help investigators reduce the number of potential suspects, increasing the efficiency of investigations [3, 8, 12, 13].

3.3 POLICIES AND REGULATIONS GOVERNING BAT RELATED TO CRIMINAL JUSTICE

The U.S. Constitution's 4th Amendment protects against unlawful search and seizure. With the advancement of modern-day technologies, this goes beyond one's person, vehicle, and real properties; these constitutional protections extend to electronic and medical data, corporate technologies, and yes, forensic data. Because of these advances in technologies, and of course, the abuses by law enforcement of obtaining DNA samples without permission or not regarding DNA collection to similar standards as, for example, entering your home and seizing evidence there, court cases and new legislation were the result, which led to the creation of rules, laws, and ethical standards for the industry.

3.3.1 Forensic Genetic Genealogy and Law Enforcement Access

With the exponential increase of ancestry testing, specifically through ancestry genealogical databases, law enforcement tools such as Investigative Genetic Genealogy (IGG) have drawn regulatory attention:

Federal Department of Justice (DOJ) Guidelines (2019) only allow FGG for violent crimes when traditional methods (like CODIS) fail, and require prosecutor approval, agency consultation, and use of databases that expressly notify users about law enforcement access [10, 14–16]. The State of Montana requires a search warrant to access consumer genealogy databases unless the user has waived privacy rights. Furthermore, the State of Maryland is more comprehensive. Maryland limits searches to violent crimes, mandates judicial authorization, informed consent from the database, licensing for labs and genealogists, plus private rights of action for privacy violations. Other states like Hawaii, Kentucky, and Wyoming restrict disclosure of genetic data to law enforcement absent a court order or user consent [15].

Law enforcement's use, and in some cases abuse, of accessing DNA data specifically from private and medical DNA databases without warrants, or by deception, has historically triggered numerous constitutional issues that include potential 4th Amendment unlawful search and seizure violations [17, 18]. As a result, case law and legislation have resulted in creating universal policies in addition to those noted above. The U.S. DOJ has developed policies and guidelines for law enforcement that are in closer alignment with those legal decisions and state and federal law [11, 19–21]. As laws change and landmark court cases address new legal challenges, it is advisable to consult current standings of case law and changes in legislation regularly to ensure legal and ethical parameters are not breached.

3.3.2 Policies of Consumer Platforms

23andMe & Ancestry generally resist law enforcement access unless legally compelled. FamilyTreeDNA allows searches by default but enables users to opt out of sharing. GEDmatch requires users to opt in for law enforcement searches; their default setting is opt in for new profiles with law enforcement [14, 21].

3.3.3 Policies of Consumer Platforms

23andMe & Ancestry generally resist law enforcement access unless legally compelled. FamilyTreeDNA allows searches by default but enables users to opt out of sharing. GEDmatch requires users to opt in for law enforcement searches; their default setting is opt in for new profiles with law enforcement [14, 21, 22]. You might want to do your personal research, checking the current policies of the various ancestry databases before selecting one for your family genealogy research, read the fine print, and check for policy updates regularly to ensure a reasonable expectation of privacy [22, 23]. Furthermore, it is advisable to research the platform's policies regarding the voluntary removal of your DNA from their database should you want to remove it in the future [16, 23–25].

From the criminal justice perspective, stay on top of any new case decisions and changes in state and federal laws regarding the access and use of DNA from commercial and medical databases. Violation of constitutional 4th Amendment search and seizure laws is serious and a distinct breach of the public trust [16, 25, 26]. There is a history of law enforcement corruption and illegal access to ancestry and medical genetic databases in the interest of criminal investigations that have resulted in ethical and legal violations [11, 19–21, 23, 27–32].

3.4 CHALLENGES AND LIMITATIONS OF BAT

Biogeographic Ancestry Testing has limitations. Forensic DNA is not a magic fix-all for automatic criminal convictions by any means. Furthermore, DNA evidence can be tainted, corrupted, or even planted at a scene to sway a criminal investigation, arrest, and prosecution. In previous chapters, you have already read some highly disturbing examples of instances where several people were wrongfully convicted, either by mistake, neglect, or intentionally corrupting the evidence. The toll on human lives is immeasurable when this happens; decades in prison, or even the execution of wrongfully convicted persons. Individuals are destroyed along with their families, all the while the guilty are walking free in society. These are atrocities equivalent to the worst war crimes. Society should expect more than simple due diligence from law enforcement, prosecutors, and defense counsel, and the standards and expectations on those involved in the criminal justice system should, in this author's opinion, be increased dramatically.

3.4.1 Accuracy and Reliability

Biogeographical ancestry testing uses single-nucleotide polymorphisms (SNPs), which are genetic markers that vary in frequency across populations. Algorithms compare a person's SNP profile with *reference datasets* from sampled populations worldwide, and these results are reported as *percentages of ancestry composition* or *regional assignments* [4, 33–35]. Some factors affecting accuracy can include the following which are found in Sections 3.4.2 to 3.4.17.

3.4.2 Reference Database Size and Diversity

Typically, larger databases with greater diversity improve the accuracy of the final determination. Underrepresented regions and populations, such as those in parts of Africa, the Indigenous Americas, and Oceania, tend to yield less accurate predictions. Populations with shared histories, such as Europeans, South Asians, and Middle Eastern groups, can be hard to accurately distinguish [1, 2, 4, 34–38].

3.4.3 Resolution Level

Continental ancestry, such as African vs. European vs. East Asian, is highly reliable with accuracy greater than 95%. *Sub-continental/regional ancestry*, such as that of distinguishing French vs. German, or West vs. East Africa, has moderate accuracy due to genetic overlap. *Fine-scale ancestry*, such as tribal or local village-level ancestry, is often unreliable because many of those populations may not be well represented in genetic databases [2–4, 6, 24, 34, 35, 37–41].

3.4.4 Algorithmic Methods

The accuracy of algorithmic methods can vary between the use of *principal component analysis (PCA), admixture modeling*, and the use of *machine learning* techniques. Accuracy can vary significantly, and the various proprietary methods are not always transparent [1, 3, 4, 24, 35, 37, 39, 40].

3.4.5 Time Depth of Ancestry

Tests typically reflect *ancestry within the last approx. 500 to 1,000 years*. Ancient ancestry beyond this timescale cannot be precisely resolved [1–4, 24, 33, 37, 39].

3.4.6 Reliability Issues & Limitations

Company-to-Company Variation: The same DNA sample may yield different ancestry estimates depending on the testing service.
Statistical Uncertainty: Confidence intervals are rarely shown in consumer reports, leading to overinterpretation.
Admixture Populations: People with recent mixed ancestry may receive variable results.
Changing Results Over Time: As companies update databases and algorithms, customer ancestry reports can shift [3, 37].

3.4.7 Overall Reliability

Highly Reliable: Broad continental ancestry assignment such as African vs. European vs. Asian.

Moderately Reliable: Regional distinctions such as Northern vs. Southern Europe.

Less Reliable: Narrow or fine-scale ethnic/tribal identifications.

Not Reliable: Claims of definitive identity, ethnicity, or legal heritage without supporting genealogical and historical evidence [1, 2, 24, 37, 38, 41].

3.4.8 Ethical Concerns

Overall considerations of individual privacy, constitutionality of access to data, medical HIPAA laws, race, politics, surveillance, and other subjects plague the plethora of situations that can give rise to privacy, security, and abuse of DNA databases. The following are some highlighted areas of concern where ethical breaches can compromise individual personal DNA forensic data.

3.4.9 Privacy and Data Security

Genetic Data Permanence: Once collected, DNA cannot truly be anonymized, and individuals may be identifiable even in "de-identified" datasets.

Third-party Sharing: Many testing companies reserve the right to share or sell data to pharmaceutical or research partners, often buried in terms of service.

Law Enforcement Access: Some databases such as GEDmatch and FamilyTreeDNA, allow or have allowed police access, raising Fourth Amendment and consent issues [18, 21–24, 42–45].

3.4.10 Informed Consent

Opaque Terms: Customers may not fully understand what they are consenting to when submitting DNA samples.

Future Use of Samples: Genetic material might be stored and used for purposes beyond ancestry testing (e.g., research and commercialization).

Familial Implications: Testing one individual reveals information about relatives who never consented [19, 23, 43, 44, 46–51].

3.4.11 Representation and Equity

Underrepresented Populations: Many global populations (Indigenous peoples, sub-Saharan Africans, Pacific Islanders) are undersampled in databases. This skew results and reinforces Eurocentric bias.

Exploitation Concerns: Indigenous and marginalized groups have raised objections to the use of their DNA without proper community consent or benefit-sharing [33, 45].

3.4.12 Accuracy, Misinterpretation, and Identity

Overinterpretation: People may conflate statistical estimates with definitive statements about ethnicity or identity.

Shifting Results: As algorithms and databases update, results can change, undermining perceived reliability.

Cultural vs. Genetic Identity: Genetic ancestry does not equate to cultural, social, or political belonging. Misuse of BAT can reinforce essentialist views of race and ethnicity [35, 38, 41, 42].

3.4.13 Discrimination and Misuse

Insurance and Employment Risks: While GINA (U.S.) protects against health insurance and employment discrimination, it does not cover life, disability, or long-term care insurance.

State Surveillance: In some jurisdictions, governments have sought access to genetic data for population monitoring or law enforcement.

Commercial Exploitation: Companies profit from consumers' genetic information, sometimes without transparent benefit-sharing [21, 23, 24, 35, 42–44].

3.4.14 Familial and Social Impacts

Unexpected Findings: Tests may reveal non-paternity, adoption, or misattributed family relationships, causing psychological harm.

Family Disputes: Inheritance, custody, or immigration cases may misuse ancestry results.

Ethnic or National Identity Claims: Misuse of results to claim membership in Indigenous or minority communities without cultural affiliation [42, 43].

3.4.15 Ethical Guidance & Standards

Professional Codes: Genetic genealogy organizations urge respect for privacy, consent, and confidentiality.
Bioethics Principles: Autonomy, beneficence, non-maleficence, and justice all apply.
Community-based Ethics: Calls for stronger **tribal sovereignty** and group consent in the use of Indigenous genetic data [21, 42–44].

3.4.16 Legal Challenges

A wide range of legal issues can result from the improper disclosure of personal DNA forensic data from corporate, medical, or even public DNA databases. The big issue across the board is privacy. Individual privacy and rights to access personal DNA records from DNA database depositories are governed by a plethora of state and federal laws, company policies, and court decisions. Below are some of the common trouble areas that present the most frequent legal challenges.

3.4.17 Ethical, Practical, and Emerging Considerations

Though This Sounds Redundant, Here Too These Considerations Apply. Many database platform consumers do not fully understand how their or their relatives' biogenetic data might be used, often learning of *a* company's data-sharing policies only after discovering them buried deep in the terms of service or upon realizing their data had been shared without their knowledge or *supposed* consent. *Companies'* policies vary as previously discussed in this chapter, thus it is imperative, for the customer concerned for their privacy, to fully research the *company's* policies and read the small print *before* submitting *one's* DNA sample to the company [23, 52].

To avoid excessive redundancy, as many of the legal challenges surrounding ethical, practical, and emerging considerations align with Section 3.4.28 Ethical Concerns, please refer above specifically regarding areas of Ethics, Privacy vs. Law Enforcement, Gene Theft and Consent, and Professional Ethical Standards. Issues and concerns surrounding BAT, and all collection, handling, and use of DNA forensic evidence, are parallel with the legal issues and challenges that result [1, 8, 11, 19–21, 23, 27, 42, 44, 45, 52–55].

3.5 BEST PRACTICES AND RECOMMENDATIONS

Biogeographical Ancestry Testing can be an especially useful tool in the criminal justice system, both in the toolboxes of the prosecution and defense alike [1, 2, 37]. Unfortunately, as we have seen from situations highlighted in previous chapters, the *human factor* of corruption exists in all venues of the criminal justice system, which has led to countless wrongful convictions, often decades old, right up to Death Row [8, 53–55]. One cannot presume the integrity of law enforcement, prosecutors, forensic scientists, or even defense attorneys [28, 29, 31, 53, 54, 56–59]. The burden of proof is on the government; however, when government agents focus on conviction numbers or allow their personal biases to influence their actions, the government can no longer be trusted. Harsh words? If you think so, ask someone who was finally exonerated after a lifetime, decades, behind bars, for crimes they did not commit. It is imperative that best practices are not only improved and followed, but in this author's opinion, that criminal penalties are harshly enforced on government agents for wrongdoing, neglect, incompetence, and especially, of course, for wanton disregard of truth when these instances occur.

3.5.1 Standards and Guidelines

Adherence to established standards and guidelines is essential for the responsible and effective use of BAT:

- **Quality Assurance Standards:** Forensic DNA laboratories using BAT must comply with quality assurance standards for forensic DNA testing and databasing, such as those recommended by SWGDAM [60, 61].
- **Accreditation:** Seeking accreditation from nationally recognized organizations like the American National Standards Institute National Accreditation Board (ANAB) or the American Association for Laboratory Accreditation (A2LA) ensures adherence to quality standards and best practices [61, 62].
- **Data Collection and Analysis Protocols:** Implementing robust data collection and analysis protocols minimizes the risk of errors and ensures the scientific validity of the findings [29, 32, 63–65].

3.5.2 Ethical Guidelines

Developing comprehensive ethical guidelines for the use of BAT in criminal investigations is imperative to ensure responsible and equitable application of the technology. These guidelines should address:

- **Protection of Vulnerable Populations:** Special attention needs to be paid to protecting vulnerable groups whose genetic data might be disproportionately impacted [20, 23, 35, 38, 42–45, 52].
- **Data Security and Confidentiality:** Strong measures must be implemented to safeguard genetic information from unauthorized access, breaches, and misuse [19, 36, 44, 66].
- **Transparent Communication and Public Education:** Openly communicating about the benefits and limitations of BAT, along with its ethical implications, is vital to fostering public trust and understanding [1, 3, 23, 35, 37, 43, 67].

3.5.3 Policy Recommendations

Policy recommendations for BAT in criminal justice include:

- **Strengthening Regulations Concerning Genetic Data Privacy:** Expanding upon existing legislation and enacting new policies to address emerging ethical challenges is crucial for adapting to the evolving landscape of genetic technologies [21–23, 35, 42, 43].
- **Promoting Diversity in Forensic Databases:** Addressing the biases in existing databases and actively working to increase the representation of diverse populations within them improves the effectiveness and fairness of BAT [35, 38, 41–43, 45].
- **Encouraging Inter-laboratory Collaboration and Standardization:** Collaborative efforts among laboratories and institutions worldwide are vital for establishing unified standards, protocols, and best practices for BAT [24, 35, 41, 42, 44].

3.6 SUMMARY

BAT is a powerful tool with the potential to significantly enhance criminal investigations and identification efforts. However, its use also raises complex challenges and legal and ethical concerns that require careful consideration. Through continuous research, development of robust methodologies, establishment of clear standards, and open dialogue about its implications, BAT can be harnessed responsibly and effectively to advance the pursuit of justice while safeguarding individual rights and privacy.

REFERENCES

[1] P. M. Schneider, B. Prainsack, and M. Kayser, "The use of forensic DNA phenotyping in predicting appearance and biogeographic ancestry," (in eng), *Deutsches Ärzteblatt International*, vol. 51–52, no. 51–52, pp. 873–880, Dec 23 2019, doi: 10.3238/arztebl.2019.0873

[2] E. Y. Y. Cheung, M. E. Gahan, and D. McNevin, "Predictive DNA analysis for biogeographical ancestry," *Australian Journal of Forensic Sciences*, vol. 50, no. 6, pp. 651–658, 2018, doi: 10.1080/00450618.2017.1422021

[3] D. McNevin, "Forensic inference of biogeographical ancestry from genotype: The genetic ancestry lab," *WIREs Forensic Science*, vol. 2, no. 2, p. e1356, 2020, doi: 10.1002/wfs2.1356

[4] H. C. L. A. T. L. Ladd, "Forensic applications of biogeographical ancestry testing: Methods and challenges," *Forensic Science International*, vol. 35, no. Jul., pp. 24–30, 2018, doi: 10.1016/j.fsigen.2018.06.009

[5] U. o. W. Florida. "Advancements in technology," Available: https://uwf.edu

[6] Researchgate. "Microhaplotypes for fine-scale ancestry inference," Available: https://researchgate.net

[7] J. Butler, *Forensic DNA Typing: Biology, Technology, and Genetics of STR Markers*, 2nd ed. Elsevier Academic Press, 2005.

[8] N. S. Udogadi et al., "Forensic DNA profiling: Autosomal short tandem repeat as a prominent marker in crime investigation," (in English), *The Malaysian Journal of Medical Sciences*, vol. 27, no. 4, pp. 22–35, 2020, doi: 10.21315/mjms2020.27.4.3

[9] T. Weng Chaing David, A. Stasi, and B. K. Dhar, "Forensic DNA profiling in the southern border provinces of Thailand: Ethical and regulatory issues," (in English), *Forensic Science International (Online)*, vol. 336, Jul, pp. 12–27, 2022, doi: 10.1016/j.forsciint.2022.111322

[10] O. M. Tuazon, R. A. Wickenheiser, R. Ansell, C. J. Guerrini, G. J. Zwenne, B. Custers, "Law enforcement use of genetic genealogy databases in criminal investigations: Nomenclature, definition and scope," *Forensic Science International: Synergy*, 2024, doi: 10.1016/j.fsisyn.2024.100460

[11] K. Lo. "Whose DNA is it anyway? Legal challenges that arise from the use of genetic genealogy in criminal investigations," Available: https://jolt.richmond.edu/2024/04/26/whose-dna-is-it-anyway-legal-challenges-that-arise-from-the-use-of-genetic-genealogy-in-criminal-investigations/

[12] P. Chowdri, "Emerging tech and law enforcement: Forensic genetic genealogy," [Online]. Available: https://www.police1.com/investigations/emerging-tech-and-law-enforcement-forensic-genetic-genealogy
[13] N. H. Lents, "Use of familial DNA in an investigation can be intrusive. But a middle ground is possible; New Jersey police's use of an infant's blood sample to ID a suspect may point to tremendous potential for solving cold cases, but it has rightfully alarmed privacy activists," *Think*. [Online]. Available: https://www.nbcnews.com/think/opinion/new-jersey-polices-use-babys-dna-undermines-publics-trust-rcna43996
[14] A. B. Association, "Regulating forensic genetic genealogy," *GPSolo eReport*, May 28, 2023. https://www.americanbar.org/groups/gpsolo/resources/ereport/archive/regulating-forensic-genetic-genealogy/
[15] A. B. Association, "Genetic genealogy and the law: Recent state legislation," *Health Lawyer*, 2024. [Online]. Available: https://www.americanbar.org/groups/health_law/resources/health-lawyer/2024-april/23andme-my-mom-my-second-cousin/
[16] U. S. D. o. Justice. "United States Department of Justice Interim Policy Forensic Genetic Genealogical Dna Analysis And Searching " *U.S. Department of Justice*. Available: https://www.justice.gov/olp/page/file/1204386/d
[17] G. H. A. C. Schuler, "A Fourth Amendment analysis of the use of genetic genealogy databases in criminal investigations," *The Journal of Criminology, Criminal Justice, Law & Society*, vol. 23, no. 2, pp. 58–75, 2018.
[18] J. A. Kennett, "Forensic genetic genealogy and the Fourth Amendment: A Fourth Amendment analysis of law enforcement access to genetic genealogy databases," *Journal of Constitutional Law - Penn Carey Law*, vol. 25, no. Jul., pp. 1475–1506, 2023.
[19] A. Kovacs, "Law enforcement access to dna databases: Privacy and legal implications," ed. The DNA Tests, 2025. https://thednatests.com/dna-test/dna-data-privacy/
[20] R. S. J. A. P. F. Williams, "The future of forensic genetic genealogy: Legal and ethical challenges," *Journal of Law and the Biosciences*, vol. 9, no. Jan., pp. 1–20, 2022, doi: 10.1093/jlb/lsab025
[21] A. Staff. "Ancestry guide for law enforcement," *Ancestry.com*. Available: https://www.ancestry.com/c/legal/lawenforcement?ranMID=50138&ranEAID=3420415&ranSiteID=Gl6mUn9YQ68-5h6X0CS3C7JHNICa28bb5Q&o_xid=0003420415&o_lid=0003420415&o_sch=Affiliate%20External
[22] A. Lee. "Can law enforcement use your ancestry DNA results? | Genetic genealogy," Available: https://www.youtube.com/watch?v=gPM7qTelfwY

[23] D. Ayres, "Loss of privacy rights: Think twice before DNA ancestry testing," *2018: Virginia Bioethics Advisory Committee*. [Online]. Available: https://www.vabioethics.com/content/2018/8/18/loss-of-privacy-rights-think-twice-before-dna-ancestry-testing
[24] M. J. B. Catherine, A. Ball, J. Byrnes, P. Carbonetto, K. G. Chahine, R. E. Curtis, J. M. Granka, E. Han, E. L. Hong, A. R. Kermany, N. M. Myres, K. Noto, J. Qi, K. Rand, D. Barry Starr, Y. Wang, L. Willmore, and A. B. Wolf, "AncestryDNA matching white paper," *Ancestrycdn.com*, 2025. [Online]. Available: https://www.ancestrycdn.com/support/us/2025/01/2025matchingwhitepaper.pdf
[25] Interim policy on forensic genetic genealogical DNA analysis and search. (2019). [Online] Available: https://www.justice.gov/opa/pr/doj-announces-interim-policy-forensic-genetic-genealogical-dna-analysis-and-search
[26] J. Kaiser, "New federal rules limit police searches of family tree DNA databases," *Science Insider*, 2019. [Online]. Available: https://www.science.org/content/article/new-federal-rules-limit-police-searches-family-tree-dna-databases
[27] J. W. Hazel and E. W. Clayton. "Law enforcement and genetic data," Available: https://www.thehastingscenter.org/briefingbook/law-enforcement-and-genetic-data
[28] M. Andrews, "Murder case of heather bogle unraveled errors and corruption in Sandusky county," [Online]. Available: https://www.facebook.com/wtol11/posts/11-investigates-revisited-the-murder-of-heather-bogle-10-years-after-her-death-w/1125825892920600/
[29] M. Weigel, "DNA evidence, forensic science and confessions that corrupt," *The Journalist Resource*, 2013. [Online]. Available: https://journalistsresource.org/criminal-justice/confessions-corrupt-evidence-cases
[30] P. DeGregory and N. Musumeci, "Dad of slain jogger 'corrupted' crime scene by cradling her body: defense," [Online]. Available: https://nypost.com/2018/11/19/dad-of-slain-jogger-corrupted-crime-scene-by-cradling-her-body-defense/
[31] K. Silva, "Corrupt Victoria Police officer jailed after falsifying evidence in drug squad investigation," [Online]. Available: https://www.msn.com/en-au/news/australia/corrupt-victoria-police-officer-jailed-after-falsifying-evidence-in-drug-squad-investigation/ar-AA1ujX4n?ocid=BingNewsVerp
[32] A. Ossola, "DNA evidence is not foolproof; believing it blindly could put the wrong people in jail," *Popular Science*, 2015. [Online]. Available: https://www.popsci.com/dna-evidence-not-foolproof/#:~:text=Juries%20and%20lawyers%20alike%20generally%20consider%20DNA%20evidence,less%20reliable%20and%20objective%20than%20most%20people%20think

[33] S. W. E. Al, "Genetic structure, divergence and admixture of Han Chinese, Japanese and Korean populations," *Genome Biology*, vol. 19, no. 1, pp. 1–12, 2018, doi: 10.1186/s41065-018-0057-5
[34] E. H. M. Jobling, T. Hurles, M. Kivisild, and C. Tyler-Smith, *Human Evolutionary Genetics*, 2nd ed. Garland Science, 2014.
[35] D. Royal, et al., "Inferring genetic ancestry: Opportunities, challenges, and implications," *American Journal of Human Genetics*, vol. 86, no. 5, pp. 661–673, 2010, doi: 10.1016/j.ajhg.2010.03.011
[36] National Institute of Standards and Technology, "Population databases for forensic DNA analysis," *National Institute of Standards*, vol. 1229, no. 1–25, pp. 1–25, 2020.
[37] C. M. Lewis, et al., "Genetic ancestry analysis on personal genomics platforms: Accuracy and limitations," *Nature Reviews Genetics*, vol. 22, pp. 507–517, 2021.
[38] S. K. Shriner, "Reconciliation of genome ancestry and self-identified race/ethnicity in African Americans," *Frontiers in Genetics*, vol. 6, p. 98, 2015, doi: 10.3389/fgene.2015.00098
[39] C. Phillips, "Forensic genetic analysis of bio-geographical ancestry," *Forensic Science International: Genetics*, vol. 18, pp. 49–65, 2015, doi: 10.1016/j.fsigen.2015.05.012
[40] M. D. S. A. R. N. Kidd, "Developing ancestral information from ancestry informative markers: Methodological considerations," *Forensic Science International.*, vol. 38, no. Mar. 2019, p. 38, 2019, doi: 10.1016/j.fsigen.2019.01.002
[41] L. Gannett, "Biogeographical ancestry and race," *Studies in History and Philosophy of Science Part C: Studies in History and Philosophy of Biological and Biomedical Sciences*, vol. 47, pp. 173–184, 2014, doi: 10.1016/j.shpsc.2014.05.017
[42] C. M. Lewis, "Ethical, legal, and social issues in ancestry testing," *Nature Reviews Genetics*, vol. 22, pp. 507–517, 2021.
[43] J. M. Butler, "Genetic data privacy: Implications of ancestry testing," *Forensic Science International: Genetics*, vol. 50, pp. 102–415, 2021.
[44] A. Kennett, "Privacy, consent and population genetics: The ethics of DNA databases," *Bioethics*, vol. 33, no. 4, pp. 422–430, 2019, doi: 10.1111/bioe.12565
[45] A. B, et al., "The ethics of genetic ancestry testing in the context of Native sovereignty," *Genetics in Medicine*, vol. 200, no. 2, pp. 393–398, 2015, doi: 10.1534/genetics.115.176461
[46] (2024). A privacy of DNA: FTC's approach to genetic data. [Online]. Available: https://www.ftc.gov/business-guidance/blog/2024/01/dna-privacy-privacy-dna

[47] E. F. Foundation. "Law enforcement access to genetic databases raises privacy concerns," *Electronic Frontier Found.* Available: https://www.eff.org/deeplinks/2022/09/law-enforcement-access-genetic-databases-raises-privacy-concerns
[48] H. L. Review, "Genetic privacy and the limitations of HIPAA," *Houston Law Review*, vol. 58, no. 2, 2021. [Online]. Available: https://houstonlawreview.org/article/36547
[49] A. C. L. Union. "Genetic genealogy and law enforcement," *American Civil Liberties Union.* Available:https://www.aclu.org/news/privacy-technology/genetic-genealogy-and-law-enforcement
[50] S. B. Elwyn, "The DNA of privacy and the privacy of DNA," *International Association of Privacy Professionals (IAPP)*, 2023.
[51] J. J. McShane, "Balancing privacy and public safety: The role of 23 and Me's DNA database in criminal investigations," [Online]. Available: https://thetruthaboutforensicscience.com/balancing-privacy-and-public-safety-the-role-of-23andmes-dna-database-in-criminal-investigations/
[52] J. G. Russell, "The ethics of DNA testing," *OnBoard 21*, vol. 1, no. 7, 2015. [Online]. Available: https://www.bcgcertification.org/skillbuilding-the-ethics-of-dna-testing. Board for Certification of Genealogists.
[53] G. M. Laporte, "Wrongful convictions and DNA exonerations: Understanding the role of forensic science," *National Institute ofe Justice Journal*, 2017. [Online]. Available: https://nij.ojp.gov/topics/articles/wrongful-convictions-and-dna-exonerations-understanding-role-forensic-science
[54] N. Elstor, "How forensic DNA evidence can lead to wrongful convictions," *Jstore Daily*, 2017. [Online]. Available: https://daily.jstor.org/forensic-dna-evidence-can-lead-wrongful-convictions
[55] D. o. S. Staff. "Causes of Wrongful Conviction," Western Michigan University, Department of Sociology. Available: https://wmich.edu/sociology/causes-wrongful-conviction
[56] L. Hamers, "When DNA Lies: The hidden flaws in forensic science," *SciTech Daily*, 2024. [Online]. Available: https://scitechdaily.com/when-dna-lies-the-hidden-flaws-in-forensic-science
[57] K. Kennett, "Forensic genealogy: Some serious concerns," *Forensic Science International: Genetics*, vol. 36, pp. 203–209, 2018.
[58] J. McShane, "DNA interpretation: The prosecutor's fallacy," [Online]. Available: https://thetruthaboutforensicscience.com/dna-interpretation/
[59] P. Gleeson, "Bombshell claims 'botched' DNA testing at Queensland forensic centre may have caused hundreds of flawed prosecutions,"

[Online]. Available: https://www.skynews.com.au/australia-news/politics/bombshell-claims-botched-dna-testing-at-queensland-forensic-centre-may-have-caused-hundreds-of-flawed-prosecutions/news-story/20c19d86511a2abd1a62e378c47a046b

[60] SWGDAM. "Quality assurance standards for forensic DNA testing," Available: https://swgdam.org

[61] A. A. f. L. A. (A2LA). "Accreditation and quality standards." Available: https://a2la.org

[62] A. N. S. I. N. A. B. (ANAB). "Accreditation for forensic DNA laboratories," Available: https://anab.org

[63] S. H. Hess Cordula, "Recovery of trace DNA on clothing: A comparison of mini-tape lifting and three other forensic evidence collection techniques.," *Journal of Forensic Sciences*, vol. 61, no. 1, 2017, doi: 10.1111/1556-4029.13246

[64] N. I. o. Justice, *What Every Law Enforcement Officer Should Know About DNA Evidence*, National Institute of Justice: National Institute of Justice, 1999.

[65] Property & Evidence Control, L. E. P. Center, Law enforcement policy center, 2021. [Online]. Available: https://www.theiacp.org/resources/policy-center-resource/property-evidence-control

[66] G. Staff. "Open-source DNA databases: What they are and how to use them," Genomelink. https://genomelink.io/blog/open-source-dna-databases

[67] D. A. Paprota, "*The Influence of Higher Education on Law Enforcement Entry Level Examination Outcomes*," Ed.D. Seton Hall University, United States -- New Jersey, 3520929, 2012. [Online]. Available: https://trident.idm.oclc.org/login?url=https://www.proquest.com/dissertations-theses/influence-higher-education-on-law-enforcement/docview/1033649479/se-2?accountid=28844, https://YV9QF3BQ4D.search.serialssolutions.com?ctx_ver=Z39.88-2004&ctx_enc=info:ofi/enc:UTF-8&rfr_id=info:sid/ProQuest+Dissertations+%26+Theses+Global&rft_val_fmt=info:ofi/fmt:kev:mtx:dissertation&rft.genre=dissertations&rft.jtitle=&rft.atitle=&rft.au=Paprota%2C+David+A.&rft.aulast=Paprota&rft.aufirst=David&rft.date=2012-01-01&rft.volume=&rft.issue=&rft.spage=&rft.isbn=978-1-267-53637-2&rft.btitle=&rft.title=The+Influence+of+Higher+Education+on+Law+Enforcement+Entry+Level+Examination+Outcomes&rft.issn=&rft_id=info:doi/

CHAPTER 4

The Accuracy of Forensic DNA Testing

Heather Miller Coyle

CONTENTS

4.1 INTRODUCTION

DNA, or deoxyribonucleic acid, is the incredible molecule of genetic inheritance. It holds the information for the creation of another individual and for the proper functioning of the human body. DNA is required as a coding molecule for functional proteins; it also has some interesting repeating sequence regions and some noncoding regions. Scientists have spent decades elucidating the function of as many regions as possible to better understand the medical needs of human culture and the basis of human genetic variation. There are typically 23 pairs of human chromosomes in a normal somatic cell, along with many circular mitochondrial

DOI: 10.4324/9781032714998-4

genomes. This chapter will address the most common forensic DNA analysis method for short tandem repeat (STR) analysis in a discussion of the collection, extraction, quantitation, and analysis steps that can be evaluated for testing accuracy in this human identification process for the autosomal chromosomes [1–5]. Ultimately, the forensic standard for human identification is high for criminal casework, paternity testing, and missing persons identification by DNA, and the United States courts mandate statistical match requirements for each of these main areas to give a weight or meaning to a DNA match.

4.2 EVIDENCE COLLECTION

The biological evidence collection step is crucial for the correct interpretation of the downstream data generated from DNA testing. The collection process is performed in a sterile, hygienic manner to avoid the collection personnel from adding their own DNA inadvertently to the sample. Sterile swabs, devices, gloves, and crime scene suits are required to contain the constantly shed cellular material from skin, as well as shed hairs, from combining with the crime scene evidence during the collection step. Otherwise, inadvertent gross DNA contamination from personnel could alter the downstream DNA results and case interpretation by resulting in mixtures. Gross contamination would result in an inaccurate result, as it adds information to the evidentiary sample that was not originally present when it was left at the scene [6]. For paternity testing, sterile buccal swabs are collected from the people involved and transported in a sterile fashion to the laboratory for DNA testing and comparisons. For missing persons casework, frequently skeletal and dental remains are processed in specialty forensic laboratories or in dedicated laboratory space to avoid cross-contamination between cases due to pulverized samples that can be aerosolized into the environment and remain on surfaces if not appropriately cleaned and maintained [7]. The evidence collection step is the most crucial to avoid contamination and maintain the integrity of the original crime scene sample to ensure accuracy in testing [8, 9].

4.3 EVIDENCE SAMPLING

Both at the crime scene and in the forensic science laboratory, it is required for the evaluator to be trained professionally to recognize biological stains and contact points where DNA deposit or transfer may have occurred on items such as bottles, cans, weapons, and clothing. Evidentiary items are highly variable. Some items can be transported from the scene as movable objects, while others cannot. Therefore, there is a decision made by the on-site examiner as to what area to collect from or whether the item is even probative to the case. This step is the highly valued evidence recognition step that determines what forms of evidence will be collected and sent to the forensic science laboratory for further examination and testing. Once at the laboratory, a scientist will examine the item further and decide if additional collection is required. Therefore, swabs from biological stains may be collected at the scene or within the laboratory and may differ on a case-by-case basis. If the probative value of the biological evidence is not recognized due to the minute size of the stain, the invisible nature of the DNA molecule, or a potential mixed stain scenario, evidence may be lost, and a full, accurate understanding of the crime scene may not always be possible. Additional rounds of evidence testing at the laboratory can be useful to clarify scene results and the number of individuals involved at the scene. There is always a chance that limited evidence processing will not provide the full information for the scene, yielding incomplete and inaccurate results, and, therefore, it is critical to assess the evidence log completely and thoroughly when deciding what items to send to the forensic laboratory.

4.4 DNA EXTRACTION

There are many formats for purification of DNA from cellular material found in fingerprints, lip prints, ear prints, from shed epithelial cells from touch contact with surfaces, and in biological stains such as blood, semen, saliva, urine, vaginal secretions, fecal matter, tears, and perspiration. These methods include phenol-chloroform extraction, Chelex resin extraction, solid-phase silica extraction, magnetic bead extraction, and FTA card extraction. Each method has benefits and limitations in achieving the purity or the quantity of DNA in recovery from a surface or item that can aid in generating a useful data result [10–13]:

- *Phenol-Chloroform Extraction*: Separates and purifies DNA by partitioning lipids, carbohydrates, etc. into the organic phase, leaving the purified DNA behind in the aqueous phase.
- *Chelex Resin Extraction*: Chelex is a binding agent for metal ions such as magnesium that removes cofactors required for DNAse enzymes to function, thus protecting DNA from degradation. DNA is released from resin by boiling into single strands.
- *Solid-Phase Silica Extraction*: Binds DNA to an inert surface and releases a washed and purified form by pH and salt concentration adjustments.

- *Magnetic Bead Extraction*: DNA binds to the magnetic beads, and the rest of the cellular material is washed away. DNA is then released and purified.
- *FTA Card Extraction*: DNA is preserved on chemically treated filter paper that lyses bacteria and viruses until released during a buffer extraction. Direct PCR from the filter is possible.

4.5 DNA QUANTITATION

The methods for DNA quantity assessment in forensic science vary and include (a) visualization with intercalating dye stains on agarose gels after electrophoresis, (b) UV-spectrophotometry, (c) higher primate DNA detection by probe binding assays, (d) real-time PCR assays, and (e) fluorometry [14, 15]. Agarose gel electrophoresis and UV-spectrophotometry are not human-specific and can yield an overestimate of DNA template quantity due to the nonspecific nature of the test. Bacteria, fungi, mold, insect, or animal DNA can all be present and mixed with the human sample to yield an overestimate of DNA yield. Crime scenes are not sterile, and these other types of organisms are commonly present and are a cause for concern with nonspecific quantitation methods.

Probe and PCR assays are more specific and are limited to the detection of higher primate and human DNA molecules to give a more accurate estimate of useful DNA yield from evidence samples. One commonly utilized forensic real-time PCR assay (Quantifiler Trio) combines data generated from a short and long target sequence from human DNA to give a more accurate assessment of usable DNA [16–18]. Since the assay requires polymerase chain reaction (PCR) amplification, it gives the user a perspective on the sample prior to the DNA profiling step by providing an estimate of the amount of DNA template that is useful to enzymatically copy. Based on the real-time PCR amplification curve, CT value (cycle threshold number), and DI (degradation index) value, the user can assess if an inhibitor is likely present that could limit the PCR process or if the DNA template is degraded into shorter fragments that could result in a lesser quantity of useful strands to use as template DNA. Both sensitivity and specificity are requirements for accuracy in the interpretation of DNA quantitation results:

- *Agarose Gel Electrophoresis*: DNA molecules are separated by size and charge in an electric field on a solid support. The DNA is exposed to dye that can then be compared to standards of known quantity for relative intensities and quantity estimates.

- *UV-Spectrophotometry*: Measurements of transmitted and absorbed light as it passes through substances. Good quality DNA will have an A260/A280 ratio between 1.7 and 2.0.
- *Probe Binding Method*: Uses DNA probes to bind to specific sequences to show quantity and specificity with a slot-blot hybridization technique.
- *Fluorometry*: Relies on molecular fluorescent dyes that preferentially bind DNA. This method compares unknown DNA samples to a standard curve of known quantities for a quantity estimate.
- *Real-Time or Quantitative PCR (qPCR) Assay*: Relies on fluorescent dye detection to quantify double-stranded DNA or specific target sequences. This method compares unknown DNA samples to a standard curve of known quantity for a quantity estimate.

4.6 PCR AMPLIFICATION

PCR is an enzymatic copying process that mimics DNA replication in the cell [19, 20]. It allows for small picogram amounts of DNA to be copied in regions of interest from the genome so that enough is produced for study and testing. The ingredients for copying DNA in a sterile tube include purified DNA template, PCR primers, magnesium chloride, free nucleotides, Taq polymerase, and buffer. The PCR components are heated and cooled at controlled temperature regimes in a thermal cycler that is specified by each STR kit manufacturer.

Since at least the early 1990s, candidate STR loci have been identified, characterized, and evaluated for their suitability for human identification multiplex PCR kits. A complete list of the forensic STR loci and research is available through the National Institute of Standards and Technology (NIST) website. The early STR amplification kits used just three or four loci to generate a DNA profile, but rapidly evolved to include six and nine loci. A core CODIS thirteen loci profile was decided upon to alleviate the possibility of coincidental matching with random individuals in the general population and genetic relatives. The core CODIS thirteen loci are: TPOX, D3S1358, FGA, D5S818, CSF1PO, D7S820, D8S1179, TH01, vWA, D13S317, D16S539, D18S51, and D21S11 [21]. Genetic relatives are expected to share a percentage of DNA in common due to shared ancestors and inheritance patterns. Shared allele values by coincidence in unrelated persons or allele sharing between genetic relatives need to be carefully considered for partial STR profile results, as coincidental matching can occur. To further alleviate this issue, the Federal Bureau of Investigation (FBI) required in January of 2017 that an additional seven loci (D1S1656, D2S441, D2S1338, D10S1248, D12S391, D19S433, and D22S1045) be added to the core CODIS loci set for uploading DNA profiles to NDIS, the National DNA Indexing System [22].

4.7 CAPILLARY ELECTROPHORESIS DETECTION

Capillary electrophoresis is the step that visualizes and records the fluorescent dye labels using electrophoresis, a method that separates DNA fragments (0–500 bases) based on size and charge in a fluidic system [23, 24]. The ability for the DNA in the sample to be detected rests on the quantity, the quality, and the injection time used to import or inject the sample into the instrument using a molecular stacking process near the injection port. The sample is injected and travels through a fluid polymer that acts as a sieving reagent to separate DNA fragments by reptation. A CCD camera captures the fluorescent dyes that were incorporated into the DNA fragments during the PCR amplification step after excitation with a laser. The CCD camera captures all fluorescence, and the mobility and shape of the DNA peaks are assessed during software analysis for accuracy in representing a true DNA fragment rather than a PCR artifact. Sampling variance in the electrokinetic injection step can yield variable or "stochastic" results for low-quantity templates. This sampling variance becomes evident with replicate testing by PCR of the same DNA extract and would be most commonly an issue with low-quantity, low-quality DNA samples, where the minor component of a DNA mixture might yield variable results, resulting in less reproducibility and accuracy.

4.8 SOFTWARE ANALYSIS

Data processing involves uploading the raw data files from the capillary electrophoresis instrument to a viewing and editing software to analyze the traces created during the detection step. These traces are commonly referenced as electropherograms or "peak profiles" that consist of a visual color image representation of what the CCD camera "sees" as the fluorescently labeled DNA fragment passes by the detector. The capillary electrophoresis instrument is sensitive to airborne particles such as dust and lint and can create artifactual disturbances in the data that can be edited out of the data using human judgment and experience. The PCR copying process can also create a series of known artifacts that also need to be edited out of DNA profiles. Forensic laboratory policy is written to describe the known artifacts and to define a true peak from an anomaly as an aid to the scientist in the data editing process. Once edited, the data can be assessed for the total number of contributors to a DNA sample using a maximum allele counting method. The locus with the greatest number of peaks is counted, and then the number of peaks is divided by two to yield the assumed number of contributors detected for the sample. Single donor samples are simple to interpret, but complex mixtures of more than two people may require additional software to biologically and mathematically model the data. The data sampling and modeling process will give the best estimate as to which donor belongs to the represented DNA profiles in the DNA mixture. These advanced software modeling programs are collectively called "probabilistic genotyping software" [25, 26].

Most scientists will agree that current human DNA testing methods and genomics have advanced human identification to the level of unique identity, except for identical twins, when full DNA profiles are present, when thirteen loci or more are tested. However, when partial DNA profiles are obtained, the level of uniqueness and ability to make an accurate identification also decreases. One might wonder how often an accidental, coincidental match would occur in a population during casework analysis. The key to this answer lies in the latest software advancements of probabilistic genotyping.

Probabilistic genotyping (PG) refers to the use of software to model data given information on the most likely number of contributors to a DNA sample after editing raw data for PCR artifacts by the analyst. Once peak heights and PCR artifacts are assessed in the modeling process, the statistical probability of a person being included or excluded from the sample can be explained by a likelihood ratio (LR). LR values range in value on a positive inclusionary scale; they also range in value on a negative exclusionary scale. Noncontributor studies identify the potential for false positive inclusions and exclusions. Most forensic laboratory validation studies use a noncontributor study to assess false inclusion or false exclusion probabilities. A noncontributor study is performed by taking human DNA profiles from a randomized population that are known not to be in the DNA mixture and comparing them to a set of known DNA mixtures and identifying how often the software registers LRs greater than (false positive) or less than (false negative) zero. Allele drop-out or partial data DNA profiles increase the risk of the software determining a false exclusion, while high degrees of allele sharing increase the risk of determining a false inclusion.

Validation study data illustrate that as the sensitivity of the PCR amplification kits and detection increases, so does the allele drop-in frequency; however, with advanced versions of STRmix™ the accuracy of detection of true contributors with a likelihood ratio statistic greater than 1000 also increased. Examples of validation study data are outlined below:

- *Erie County Central Police Services Forensic Laboratory Validation*: [27]
  - STRmix™ software version 2.3 (2015)
  - Validation study (Identifiler Plus kit, 29 cycles, 60 RFU AT)
  - Drop-in frequency 0.0019
  - False positive rate not available
- *New York City Office of the Chief Medical Examiner Forensic Laboratory Validation*: [28]
  - STRmix™ software version 2.4.05 (2016, updated version 2.4.08, 2019)
  - Validation study (Fusion kit, 29 cycles, 50 RFU AT)
  - Drop-in frequency 0.0024
  - For 2–3 person mixtures: false positive rate 0.5865% LR >1; 0.0015% LR > 1000
- *New York City Office of the Chief Medical Examiner Forensic Laboratory Validation*: [29]
  - STRmix™ software version 2.7 (2021)
  - Validation study [Fusion 5C kit, 29 cycles, ATs of 85 RFU (Blue), 120 RFU (Green), 130 RFU (Yellow), and 160 RFU (Red)]
  - Drop-in frequency 0.0087
  - For 2-3-4 person mixtures: false positive rate 0.6467% LR > 1; 0.0001% LR > 1000

4.9 ACCURACY IN DNA IDENTIFICATION USING THE STR METHOD

The term "accuracy" refers to the correctness of the answer. It differs from reliability and reproducibility in that the answer is precise and correct rather than consistent or duplicatable. There are many steps taken in the forensic DNA profiling process. All the steps must be performed correctly for accuracy in forensic analysis. There are some components of the testing process that can be identified as common places where errors can occur. Some of these areas have already been discussed in this chapter, but a few additional areas stand out as significant.

One of these areas includes PCR amplification efficiency, which can affect quantity estimates and the quality of the electrophoretic traces. It is commonly known that a low quantity (<100 pg) DNA template can produce stochastic DNA results that are not reproducible from test to test. Stochastic results can call into question the accuracy of the result if the data is not reproducible. This is one of the main criticisms of low copy number (LCN) DNA testing in that replication does not enhance or improve consistency in the PCR results from the same sample, no matter how many times the testing is performed [30]. In LCN test methods, "stray" or nonreproducible allele detection is common, and specialized interpretation strategies were used to score reproducible alleles for a profile while discarding nonreproducible but still detectable alleles. This strategy results in using the reproducible portion of the data set, but can confound the analyst in determining the number of contributors detected in the DNA sample. Enhanced PCR cycling conditions can also enhance the detection of contaminant alleles.

PCR amplification efficiency is controlled in forensic laboratory policy by determining the appropriate quantity of DNA template to amplify, with most United States forensic science laboratories having a lower limit assigned below which the sample can no longer be accurately tested [31]. PG software also measures PCR per locus amplification efficiency as a mass parameter in the assessment of peak height, indicating it is an important feature of determining the accuracy of a DNA profile in the computer modeling of data. The other important mass parameters are the initial template quantity and the level of degradation of the DNA template or the quality of the starting template for PCR.

4.10 ACCURACY IN MITOCHONDRIAL DNA SEQUENCING

Mitochondrial DNA, often abbreviated mtDNA, is a circular, double-stranded molecule found in abundance in most somatic cells. It is in the mitochondria, organelles found in cellular cytoplasm, that aid in cell respiration and metabolism. The mtDNA genome is approximately 16,500 base pairs and encodes 37 genes needed for normal mitochondrial processes. The genetic inheritance of the mtDNA molecule is useful in forensic science since it is passed through familial generations from mother to offspring and can be used to verify genetic lineages. The mtDNA genome has been studied as variants or mutations can lead to medical disorders including diabetes, kidney failure, muscle weakness, heart disease, and Alzheimer's disease.

DNA sequencing is a method for reading the order of the nucleotide bases on the mtDNA molecule. Several different versions of DNA sequencing exist, including Sanger sequencing, Maxam-Gilbert sequencing, pyrosequencing, and next-generation sequencing (NGS)/massive parallel sequencing (MPS):

- *Sanger Sequencing*: This method is also called chain termination sequencing and is considered the "gold standard" as it has high accuracy. Dideoxynucleotides with fluorescent dye labels terminate DNA chain elongation during DNA synthesis, and then capillary electrophoresis is used to separate and detect each fragment by size. This method is for single-source samples.

- *Maxam-Gilbert Sequencing*: This is a chemical modification method for specific bases in DNA, followed by cleavage of the DNA backbone to determine the sequence. This method is for single-source samples.
- *Pyrosequencing*: This method determines the release of pyrophosphate during DNA synthesis in real time. This method is also for single-source samples.
- *Next-Generation Sequencing*: Clustering and sequencing by synthesis are used to simultaneously read millions of DNA fragments at the same time. This method uses bioinformatics to reassemble the fragments into whole genomes or targeted regions of DNA. This method is also called massive parallel sequencing. This method is advantageous for use with DNA mixtures.

For DNA sequencing, the accuracy may vary based on the method used; however, in general, the accuracy level is 99% or higher for achieving the correct identification of the nucleotide bases in the target sequence for the Sanger sequencing method. NGS accuracy varies with error rates between 0.01% and 15% and this is the trade-off between the traditional Sanger sequencing technology and the more rapid, higher output sequencing methods. Accuracy is measured by factors such as read accuracy (machine error/base ambiguity in software analysis), consensus accuracy (average of multiple reads to reduce random error), and quality scoring (confidence value for accuracy of the read) [32, 33]. Accuracy is influenced by the quality and quantity of the starting DNA template, the coverage rate (a greater number of reads increases accuracy), and the sequencing method (each method has a varying error rate). The accuracy may be reported to the user using the Phred quality score or the percentage of correct bases scored per total number of bases read. Mitochondrial DNA is a useful adjunct to autosomal DNA tests, but it is not as discriminating a test method as some individuals share maternal haplotypes by chance.

4.11 ACCURACY IN Y-STR TEST METHOD

Y-STR technology is like autosomal STR analysis except for the targeting of genetic loci that reside on the male Y chromosome. The Y chromosome is paternally inherited and passed from father to son to connect family lineages. It is useful in forensic DNA analysis for criminal sexual assault casework, missing persons casework, and for other forensic samples that have high concentrations of female DNA combined with low quantities of male DNA. The Y-STR mutation rate (approximately 0.002% or 2 mutations per locus per 1000 generations) is higher than for most autosomal STR loci, and this can be useful for certain applications such as confirmation of forensic genetic genealogy [34]. Rapidly mutating Y-STR loci (approximately 0.01% or 1 mutation per locus per every 100 generations) may also be useful to distinguish between close genetic relatives when a mutation has occurred between a parental and offspring generation [35]. Y-STR testing is not as discriminating as autosomal STR testing, and some paternal lineage haplotypes can be shared in the population by coincidence.

4.12 ACCURACY IN THE SINGLE-NUCLEOTIDE POLYMORPHISM (SNP) TEST METHOD

Single-nucleotide polymorphisms (SNPs) are single-base differences that can occur in coding or noncoding regions of DNA, and if analyzed in sufficient quantity, can be used to identify an individual. Typically, SNP analysis is used in anthropology and forensic science casework to support inferences of biogeographical ancestry (BGA), prediction of physical features such as eye color, hair color, and skin tone (forensic DNA phenotyping or FDP), and shared familial genetics for investigative forensic genetic genealogy (IFGG) [36, 37]. It is difficult to determine an accurate rate for the SNP method in general, as the rate varies based on the SNP in question. SNP chip analyses use 500,000 or more SNPs to analyze samples and are considered highly accurate, with prediction values for common variants at 99% or higher. SNP tests, however, are not very accurate for rare variant detection, and most SNP tests require that the SNP variation be present in at least 1% of the population. SNP tests for familial search identify individuals that share large segments of DNA, and many can accurately identify individuals to the second, third, and fourth cousin levels to provide investigative leads in criminal cases. This method is paired with searchable databases generated by the general population seeking information on ancestry that has provided consent for use by law enforcement for cold case investigations. Not all databases are searchable, but GedMatch Pro, DNASolves, and FamilyTreeDNA have been used for law enforcement investigations. SNP test methods for the purpose of providing investigative leads have been successful but are also controversial, as genetic privacy issues have been raised by donors of the DNA samples. These methods are currently used in academic research and by private forensic companies that are in the process of being accredited to forensic DNA standards.

4.13 SUMMARY

Accuracy in forensic DNA analysis is imperative, as in criminal casework, the scientist strives to identify the correct answer. The consequences of inaccuracy are great, as the wrong individual may suffer in the courts, and incorrect prosecution could result if faulty test results are reported. For this reason, quality assurance programs and forensic science laboratory accreditation programs are in place in most DNA testing laboratories as a monitoring system to identify and document systematic errors in testing, test methods, and equipment. In addition, both developmental and internal validation studies are required by manufacturers and users to assess the validity of new test methods prior to use in forensic casework. The data from these programs and studies are continuously evaluated by laboratory management and the courts, as evidenced by admissibility hearings and legal decisions authorizing their use under different casework circumstances. Over time, some forensic DNA methods have been wholly embraced by the forensic science community as gold standards, while others have been rejected based on their reduced accuracy and error rates.

REFERENCES

1. Tagliabracci A, Buscemi L, Sassaroli C, Paoli M, Rodriguez D. Allele typing of short tandem repeats by capillary electrophoresis. *Int J Legal Med.* 1999;113(1):26–32.
2. Martín P, de Simón LF, Luque G, Farfán MJ, Alonso A. Improving DNA data exchange: validation studies on a single 6 dye STR kit with 24 loci. *Forensic Sci Int Genet.* 2014 Nov;13:68–78.
3. Liu F, Jia F, Sun F, Zhao B, Shen H. Validation of a multiplex amplification system of 19 autosomal STRs and 27 Y-STRs. *Forensic Sci Res.* 2019 Nov 5;5(4):292–299.
4. Gill P, Koumi P, Allen H. Sizing short tandem repeat alleles in capillary array gel electrophoresis instruments. *Electrophoresis.* 2001 Aug;22(13):2670–2678.
5. Lazaruk K, Walsh PS, Oaks F, Gilbert D, Rosenblum BB, Menchen S, Scheibler D, Wenz HM, Holt C, Wallin J. Genotyping of forensic short tandem repeat (STR) systems based on sizing precision in a capillary electrophoresis instrument. *Electrophoresis.* 1998 Jan;19(1):86–93.
6. Goray M, Mitchell JR, van Oorschot RA. Evaluation of multiple transfer of DNA using mock case scenarios. *Leg Med (Tokyo).* 2012 Jan;14(1):40–46.

7. Ge J, Budowle B, Chakraborty R. Choosing relatives for DNA identification of missing persons. *J Forensic Sci.* 2011 Jan;56(Suppl 1):S23–S28.
8. Lee HC, Ladd C. Preservation and collection of biological evidence. *Croat Med J.* 2001 Jun;42(3):225–228.
9. Plaza DT, Mealy JL, Lane JN, Parsons MN, Bathrick AS, Slack DP. Nondestructive biological evidence collection with alternative swabs and adhesive lifters. *J Forensic Sci.* 2016 Mar;61(2):485–488.
10. Nimbkar PH, Bhatt V. A review on touch DNA collection, extraction, amplification, analysis and determination of phenotype. *Forensic Sci Int.* 2022 Jul;336:111352.
11. Finaughty C, Heathfield LJ, Kemp V, Márquez-Grant N. Forensic DNA extraction methods for human hard tissue: A systematic literature review and meta-analysis of technologies and sample type. *Forensic Sci Int Genet.* 2023 Mar;63:102818.
12. Sessa F, Pomara C, Esposito M, Grassi P, Cocimano G, Salerno M. Indirect DNA Transfer and Forensic Implications: A Literature Review. *Genes (Basel).* 2023 Nov 28;14(12):2153.
13. Stray JE, Liu JY, Brevnov MG, Shewale JG. Extraction of DNA from Forensic Biological Samples for Genotyping. *Forensic Sci Rev.* 2010 Jul;22(2):159–175.
14. Lee SB, McCord B, Buel E. Advances in forensic DNA quantification: a review. *Electrophoresis.* 2014 Nov;35(21-22):3044–3052.
15. Jäger R. Genomic multicopy loci targeted by current forensic quantitative PCR assays. *Genes (Basel).* 2024 Oct 5;15(10):1299.
16. Knight KL, Mauriello A, Williams G. Quantitation of DNA using the applied biosystems quantifiler trio DNA quantification kit. *Methods Mol Biol.* 2023;2685:175–188.
17. Liu JY. Direct qPCR quantification using the Quantifiler Trio DNA quantification kit. *Forensic Sci Int Genet.* 2014 Nov;13:10–19.
18. Holt A, Wootton SC, Mulero JJ, Brzoska PM, Langit E, Green RL. Developmental validation of the Quantifiler HP and Trio kits for human DNA quantification in forensic samples. *Forensic Sci Int Genet.* 2016 Mar;21:145–157.
19. Zhu H, Zhang H, Xu Y, Laššáková S, Korabečná M, Neužil P. PCR past, present and future. *Biotechniques.* 2020 Oct;69(4):317–325.
20. McDonald C, Taylor D, Linacre A. PCR in forensic science: A critical review. *Genes (Basel).* 2024 Mar 29;15(4):438.
21. Budowle B, Shea B, Niezgoda S, Chakraborty RJ. CODIS STR loci data from 41 sample populations. *Forensic Sci.* 2001 May;46(3):453–489.
22. Moretti TR, Moreno LI, Smerick JB, Pignone ML, Hizon R, Buckleton JS, Bright JA, Onorato AJ. Population data on the expanded CODIS

core STR loci for eleven populations of significance for forensic DNA analyses in the United States. *Forensic Sci Int Genet.* 2016 Nov;25:175–181.
23. Butler JM, Buel E, Crivellente F, McCord BR. Forensic DNA typing by capillary electrophoresis using the ABI Prism 310 and 3100 genetic analyzers for STR analysis. *Electrophoresis.* 2004 Jun;25(10-11):1397–1412.
24. Shewale JG, Qi L, Calandro LM. Principles, practice, and evolution of capillary electrophoresis as a tool for forensic DNA analysis. *Forensic Sci Rev.* 2012 Jul;24(2):79–100.
25. Coble MD, Bright JA. Probabilistic genotyping software: An overview. *Forensic Sci Int Genet.* 2019 Jan;38:219–224.
26. Buckleton JS, Bright JA, Gittelson S, Moretti TR, Onorato AJ, Bieber FR, Budowle B, Taylor DA. The probabilistic genotyping software STRmix: Utility and evidence for its validity. *J Forensic Sci.* 2019 Mar;64(2):393–405.
27. Erie county central police services forensic laboratory validation: https://johnbuckleton.wordpress.com/wp-content/uploads/2016/09/strmix-implementation-and-internal-validation-erie-id-plus.pdf
28. New York City Office of the Chief Medical Examiner Forensic Laboratory Validation. Internal Validation of STRmix™ V2.4 for Fusion NYC OCMEE. OCME STRmix™ Internal Validation 18 November 2016. Updated 20 December 2019. https://www.bers.nyc.gov/assets/ocme/downloads/pdf/STRmix-V2-4-Fusion-5C-Validation%20Summary.pdf
29. New York City Office of the Chief Medical Examiner Forensic Laboratory Validation. NYC OCME Internal Validation of STRmix™ v2.7 for Fusion 5C/3500xL Data - STRmix™ Pa Parameters. Department of Forensic Biology. Charles S. Hirsch Center for Forensic Sciences. https://www.nyc.gov/assets/ocme/downloads/pdf/strmix_v2_7_parameters.pdf
30. Schulz R, Terry SF. The science, applications, and ethical concerns surrounding low copy number DNA analysis. *Genet Test Mol Biomarkers.* 2015 Jun;19(6):281–282.
31. Ruijter JM, Barnewall RJ, Marsh IB, Szentirmay AN, Quinn JC, van Houdt R, Gunst QD, van den Hoff MJB. Efficiency correction is required for accurate quantitative PCR analysis and reporting. *Clin Chem.* 2021 Jun 1;67(6):829–842.
32. Macken WL, Falabella M, Pizzamiglio C, Woodward CE, Scotchman E, Chitty LS, Polke JM, Bugiardini E, Hanna MG, Vandrovcova J, Chandler N, Labrum R, Pitceathly RDS. Enhanced mitochondrial genome analysis: bioinformatic and long-read sequencing advances and their diagnostic implications. *Expert Rev Mol Diagn.* 2023 Jul–Dec;23(9):797–814.

33. Alkanaq AN, Hamanaka K, Sekiguchi F, Taguri M, Takata A, Miyake N, Miyatake S, Mizuguchi T, Matsumoto N. Comparison of mitochondrial DNA variants detection using short- and long-read sequencing. *J Hum Genet.* 2019 Nov;64(11):1107–1116.
34. Dupuy BM, Stenersen M, Egeland T, Olaisen B. Y-chromosomal microsatellite mutation rates: differences in mutation rate between and within loci. *Hum Mutat.* 2004 Feb;23(2):117–124.
35. Boattini A, Sarno S, Bini C, Pesci V, Barbieri C, De Fanti S, Quagliariello A, Pagani L, Ayub Q, Ferri G, Pettener D, Luiselli D, Pelotti S. Mutation rates and discriminating power for 13 rapidly-mutating Y-STRs between related and unrelated individuals. *PLoS One.* 2016 Nov 1;11(11):e0165678.
36. Kayser M, Branicki W, Parson W, Phillips C. Recent advances in forensic DNA phenotyping of appearance, ancestry and age. *Forensic Sci Int Genet.* 2023 Jul;65:102870.
37. Pontes L, Sousa JC, Medeiros R. SNPs and STRs in forensic medicine. A strategy for kinship evaluation. *Arch Med Sadowej Kryminol.* 2017;67(3):226–240.

CHAPTER 5

Why Forensic Investigative Genetic Genealogy Is Important

Steven Smugeresky

CONTENTS

5.1 INTRODUCTION

Why is FIGG important? Imagine you are a homicide and sex crimes detective. You are called out to the scene of a violent crime. You and your crime scene technicians work the scene, gathering, cataloging, and preserving the evidence. You and your fellow detectives find and interview potential witnesses. The case is worked diligently, but unfortunately, nothing leads

DOI: 10.4324/9781032714998-5

you to a potential suspect. You have excellent latent fingerprints from the scene, but there are no matches in RAFIS. You have great DNA from the scene, but there are no matches in CODIS. The investigation stalls, and, before you know it, the leads dry up, and the case becomes cold.

Now imagine you are the victim of a sexual assault, or the close family member of a homicide victim, and the investigator comes to meet with you to explain how the case is proceeding. The investigator empathetically explains that leads were followed but led nowhere; latent fingerprints were placed in the system, but no matches were found; and the DNA collected is good and usable, but, just like the fingerprints, there was no match. The investigator explains that the fingerprints and DNA will stay in the system until they are identified, but that, of course, provides little consolation. Days become weeks, weeks become months, months become years, and years become decades with no news, no developments, and no matches.

Now imagine if you could take that DNA, which was so good from the crime scene, and place it in a database, and that database could spit out hundreds, if not thousands, of people that share large or small amounts of DNA with your suspect. Then imagine that for each amount of DNA shared, you could also figure out how your suspect may be related to the person who shares DNA. In a few weeks or months of genealogy work, you find your potential suspect, and what seemed to be unsolvable is solved. That, in a nutshell, outlines Forensic Investigative Genetic Genealogy (FIGG) capabilities, and it's a real game changer for Major Criminal Case investigations. FIGG, in certain cases, is able to solve the unsolvable. That is why it's important. In this chapter, you will read brief case studies of solved FIGG cases, and the summarized reflections of persons involved—and the human impact—to illustrate the importance of FIGG.

5.2 THE GERMANTOWN ELDERLY RAPIST

In August of 2007, an unknown male started a series of attacks that heightened in 2010 through 2012 in the Germantown and Gaithersburg areas of Montgomery County, Maryland. The first attack started when the victim was awakened by an unknown male intruder in her bedroom. The intruder had his pants down, and his penis was exposed. He threatened the victim that he had a handgun, saying that he would shoot her if she didn't do as he said. He demanded the victim perform fellatio on him. The victim struggled and eventually fought the assailant off. The suspect fled the scene and dropped his t-shirt behind. This shirt was logged and placed into evidence. Despite the work of detectives and support units, this case would grow cold because of the lack of new leads. This case would stay cold until 2017, when cold case detectives reexamined the case and placed additional items for DNA testing, one being the t-shirt left behind by the suspect. Detectives soon learned that the DNA obtained from this t-shirt matched three other unsolved rapes in the Germantown area. These rapes were the work of a suspect dubbed the Germantown Elderly Rapist. This suspect preyed on the most vulnerable of persons in society, elderly women.

On June 19th, 2010, around 3:00 am, an unknown male intruder entered an apartment in Germantown, Maryland, through an unlocked front door and encountered a 68-year-old woman in her bed. The suspect held an unknown weapon to the victim's head and forcibly raped her. The suspect fled the scene into the darkness undetected. Patrol units and investigators responded and circulated, but no suspect was found. DNA evidence was recovered from a SANE (Sexual Assault Nurse Examiner) kit that was performed with the victim. There was no match in CODIS.

In August of 2010, an unknown intruder crawled through a first-floor window in the early hours of the morning at a senior living center in Germantown, Maryland. The suspect was armed with a pair of scissors. The suspect found an 86-year-old woman resident of the center, hit her several times, and raped her. The suspect would once again flee the scene undetected while police units searched and stopped anyone in the area. DNA evidence would be recovered from a SANE kit performed with the victim. This evidence would match a case-to-case hit with the June 19th, 2010, rape.

In January of 2011, the same suspect returned to the victim's residence of the June 2010 rape. He entered through a window and proceeded to the victim's bedroom, where he raped her a second time. The suspect would disappear into the darkness just like in the previous incidents. DNA evidence was recovered from the crime scene and placed into CODIS, resulting in another case-to-case hit with the rapes from June and August 2010. These cases were worked diligently but would gradually grow cold until 2017 when these cases were linked to the 2007 sex assault. The detectives who received this 2017 hit worked the case and decided to try a new DNA technology in 2018 called FIGG. In September of 2018, a suspect was developed through the use of FIGG. A deep dive investigation into this suspect revealed enough information to obtain a search warrant for his DNA. The search warrant was served, and the suspect's DNA was collected. The suspect's DNA was tested and compared to the crime scene DNA from the four cases and matched. The suspect was arrested and charged with these four cases in September 2018. This suspect pleaded guilty in February 2019 and was sentenced to two life sentences in May of 2019.

5.2.1 Case Study Reflections

The 86-year-old rape victim from the August 2010 rape wrote a poem that was published in the *Washington Post*. This victim entitled her words as *HELL*. In this poem, she says that she never believed in Hell, the opposite of heaven, or the opposite of love. She had never experienced these feelings in her 86 years of life until this violent intruder forced them upon her. She goes on to describe her attacker as a dehumanizing being. The 69-year-old victim of the June 2010 & February 2011 rapes expressed anger and determination not to let this violent offender get away with his heinous acts in an article published by the Germantown, Maryland *Patch*.

True tragedy is to live a life the way you know it, and then suddenly—with a senseless violent act of a complete stranger—you are thrown into the worst feelings and emotions imaginable while your offender doesn't think twice. If FIGG as a tool can assist in taking these offenders off the street so that fewer people experience their heinous deeds, let's use it.

5.3 THE ANGIE DODGE MURDER

Note: This case was selected for the unique FIGG usage. The writer has written this case study based on his interpretation of press and media releases of the events and not personal involvement.

In June of 1996, a young woman named Angie Dodge was raped and murdered in her Idaho Falls apartment. Investigators worked diligently to develop leads. Months after Dodge's murder, investigators began to focus on a suspect by the name of Christopher Tapp. Tapp was interviewed by investigators at length and eventually confessed to being involved in Angie Dodge's murder. Tapp was arrested on the strength of his confession, even though (a) his DNA didn't match the DNA evidence left behind at the crime scene and (b) no other physical evidence could be linked to him. Throughout the trial, Tapp's confession outweighed the lack of DNA and physical evidence that could be linked to him. This point was argued, and eventually explained, that multiple people participated in the murder, which is the reason for other DNA on the scene. Tapp was convicted in 1998 and sentenced to 30 years in prison.

In 2007 the Idaho Innocence Project began looking into Christopher Tapp's confession and conviction. In 2008, Mrs. Dodge was still searching for more answers in her daughter's killing. More than a decade after the murder, investigators still had not identified any other suspects. It was pointed out at Tapp's trial that the theory was that multiple individuals participated in Angie's murder. The Dodge family continued through the years to seek justice for their daughter and push for the unknown DNA to be identified. As evidence mounted of Tapp's potential innocence, the Dodge family began to entertain the thought that the wrong person was in prison for their daughter's murder. Advances in DNA testing eventually led to more of the case's evidence being tested and retested. The Idaho Innocence Project worked tirelessly on Tapp's case, accumulating evidence. In 2017, Tapp's rape conviction was vacated, and his murder sentence was reduced to time served. In 2018, a lead on a new suspect was developed through the use of FIGG. This lead led to the identification of Brian Leigh Dipps, Sr. Investigators obtained Dipps' DNA, had it tested, and it matched the DNA evidence left behind at Angie Dodge's murder scene. Dipps was arrested and confessed to investigators a short time later.

5.3.1 Case Study Reflections

Tapp's Lawyer expressed that FIGG will be significant in the future, identifying new suspects in incidents where the wrong person was convicted. CeCe Moore, who worked on this case and is the lead genealogist at Parabon Nanolabs in Virginia, said that the use of FIGG to assist in identifying suspects in criminal cases—and exonerating the innocent—isn't going away anytime soon. This case study is highlighted to show the perseverance and push of the victim's family to ensure justice and seek answers. It also illustrates that FIGG is not just used to convict suspects but also to set the innocent free and clear the name of those wrongly accused.

5.4 THE MURDER OF DR. LE BICH-THUY BY A SERIAL RAPIST

On July 28th, 1989, a woman left her work on Executive Blvd. in Rockville, Maryland. She took a bus from Executive Blvd. to the Twinbrook Metro, where she caught another bus to Edmonston Dr. close to Rockville Pike in Rockville. It was around 10:10 pm when this woman walked from the bus toward her residence. Not too long after she began to walk, she was grabbed from behind and dragged into the bushes, where she was choked and violently raped. The suspect fled the scene, and the victim was able to go to a nearby residence to call the police. Evidence was obtained from an examination, and a DNA profile was developed at a later date.

On October 3, 1994, Patrol officers from the Rockville City Police Department responded to 1639 Martha Terrace, Rockville, MD,. to check on the welfare of Doctor Le Bich-Thuy, who was the resident. A coworker of Dr. Bich-Thuy's at Children's Hospital in DC called and requested the welfare check because Dr. Bich-Thuy had not been seen at work for several days. Officers were dispatched to Dr. Bich-Thuy's house to check her welfare. When officers arrived, they found dried blood on the front porch of the residence. Officers continued to search and found Dr. Bich-Thuy deceased in her backyard. Dr. Bich-Thuy had been raped, strangled, and bludgeoned to death with what investigators believed to be a large cement landscaping stone found near her body. Investigators would find that Dr. Bich-Thuy's mailman had discovered her house keys next to the pool of blood on the front stoop of her residence on September 29, 1994. Dr. Bich-Thuy's Metro fare card showed that she left the Twinbrook Metro Station on September 28, 1994, sometime after 1930 hours. This information would help the investigators narrow down the timeframe of Dr. Bich-Thuy's attack to when she arrived home from work on the evening of September 28th. DNA evidence was recovered during Dr. Bich-Thuy's autopsy and was tested. This DNA matched the suspect's DNA in the earlier rape from 1989.

In early 2019, FIGG was utilized to build family trees and develop a lead that would identify a suspect in both Dr. Bich-Thuy's murder and the 1989 rape. Investigators would soon find out that the suspect was deceased. This suspect died of a drug overdose in 2017 in West Virginia. Detectives consulted with the Office of the Chief Medical Examiner (OCME) in West Virginia and were able to receive blood evidence that was taken from the suspect at the time of his death. Investigators submitted this blood evidence to the Montgomery County Crime Lab for DNA testing. The DNA developed was a match to Dr. Bich-Thuy's murder and the suspect of the 1989 rape.

After the Identification of the suspect from Dr. Bich-Thuy's murder, research into other possible cases from the area—and that fit the timeframe—was looked at to see if the suspect may be responsible for more attacks. One of the cases that was uncovered was found in a crime watch article from the summer of 1989, which listed the earlier 1989 rape, and another rape close in proximity and only days apart from the first one. The modus operandi (MO) and descriptions were similar. On July 26, 1989, a young woman was walking home in the area of 500 College Parkway in Rockville when she was grabbed from behind, beaten, knocked unconscious, and raped by an unknown attacker. A sex assault kit was obtained and placed into evidence. In 2020, this DNA evidence was submitted to the Montgomery County Crime Lab and retested to be compared to the DNA of Dr. Bich-Thuy's murder. This DNA matched Dr. Bich-Thuy's murder.

5.4.1 Case Study Reflections

Henry Plotkin, a coworker and close friend of Dr. Bich-Thuy, was interviewed after the suspect in his friend's murder was identified for a true crime documentary called Bloodline Detective: *Rockville Horror*.

Mr. Plotkin complimented Dr. Bich-Thuy's beauty and superb intelligence. He goes on to describe her tenacity which might even be recalled as feisty when working on her research. Dr. Bich-Thuy did a significant amount of research pertaining to AIDS.

Dr. Bich-Thuy's murder did not just rob her of her life way too soon, but it also robbed society of her potential research and medical contributions. Even though the suspect in this case was deceased upon his identification, it allowed loved ones of the victim and the investigators closure and answers that were sought after for decades.

5.5 JOHN DOE 2000

Note: This case was selected for the unique FIGG usage. The writer has written this case study based on his interpretation of press and media releases of the events and not personal involvement.

On February 14th, 2000, the remains of a John Doe were found in Weld County, Colorado, in the area of 8th St. and Weld County Road 34. The remains were of a male in an advanced state of decomposition. Investigators worked with an anthropologist to determine as much information about the male as possible, to try and obtain information that could potentially identify him. Investigators and medical examiners found no obvious evidence of foul play or trauma, and the manner of death could not be determined due to the advanced stage of decomposition. This case quickly ran cold as this male became known as *John Doe 2000*. Twenty-two years after John Doe 2000 was found, investigators tried a new forensic technology in an attempt to generate new leads to identify him. FIGG was used, and in 2024, investigators located potential family members of John Doe 2000 who volunteered their DNA. DNA tests were done, and in December 2023, John Doe 2000 was tentatively identified as Christopher Scott Case; further DNA testing confirmed this identification.

5.5.1 Case Study Reflections

Detective Byron Kastilahn, the lead investigator on the case of John Doe 2000, noted that without FIGG, Mr. Case would never have been identified. Detective Kastilahn is excited and motivated to move forward using FIGG on other UHR cases in his jurisdiction. He hopes to bring answers to more families in unidentified human remains (UHR) cases. The Weld County Sheriff's Office expressed that, regardless of the passage of time, their department has a dedicated pursuit to generate investigative leads through the use of new forensic advancements.

UHR cases that aren't tied to suspicious circumstances or homicides often don't draw the amount of media attention that they should. FIGG should be used as much as possible in UHR cases. Every deceased person has the right to be identified properly, and their loved ones, who may have been looking for answers, can finally get them. FIGG is not just being used in UHR cases by law enforcement, but throughout the world to identify the remains of disaster victims and unidentified war victim remains. The number of applications for FIGG, in a variety of identification contexts, continues to grow.

5.6 THE POTOMAC RIVER RAPIST

Throughout the 1990s, a serial rapist and murderer dubbed the Potomac River Rapist terrorized Montgomery County, Maryland, and the District of Columbia. This suspect was connected to eight DNA-related rapes and one murder, while being suspected of multiple other rapes throughout the DC Metro area. His attacks began on May 6th, 1991, in Gaithersburg, Maryland. He attacked three more women in 1991, another in 1992, another in 1994, another in 1996, and two more in 1997. This rapist stalked his victims and attacked them in their own homes. At least one of these victims was raped while her children were in the next room, another was a teenage babysitter, and another was a housekeeper. In 1998, this rapist then escalated his attacks, perpetrating his first murder. On August 1, 1998, a 28-year-old biology intern on a fellowship in DC named Christine Mirzayan went to a barbecue at a fellow intern's home. Mirzayan was highly regarded and a rising star in the biology field, having a passion for science policy. She had a PHD in biochemistry and had studied at Yale. She vanished while walking home from the barbecue. Her body was later found in a DC Park. Mirzayan had been savagely raped and brutally beaten to death. DNA from the incident was linked to cases related to the Potomac River Rapist.

In 2019, a lead was developed using FIGG to ultimately identify the Potomac River Rapist. DNA was collected from the suspect, tested, and matched to DNA evidence left behind by the Potomac River Rapist at each of his crime scenes. The Potomac River Rapist was located and arrested in Conway, South Carolina. After almost three years of trial prep and court motions, the Potomac River Rapist was set to go to trial. The weekend before this man would go to court, to be held accountable for his crimes, he took his own life.

5.6.1 Case Study Reflections

Christine had a passion for living life to its fullest. She was passionate about her work and the possibilities it could bring to solve world problems through science. She was taken too early in life in a senseless act of violence by a man who, all the way to his death, never showed any ounce of remorse for his victims. Christine's family, and the victims of this man's brutal rapes, were motivated to see him stand trial and take responsibility for his senseless and violent actions. Victims expressed that the trauma from their attack will never allow them to get their attacker's voice out of their heads. They believed that seeing him in court and hearing a verdict of guilt might help bring them a little closure in some form. But with the last-ditch effort to attempt to keep control over his victims, this man decided to take his own life—robbing Christine's family, his rape victims, the prosecutors, investigators, and all the people he left in the wake of his violence, the chance to see justice in the courtroom. One of his rape victims, after his death, expressed her anger and frustration toward his last act of selfishness. She said that closure will not be easy and that, for the victims, his taking his own life will not be the same as if he had stood trial, been convicted, and been jailed for all of his crimes. She also expressed empathy for Christine's family and the other victims—in addition to showing compassion for this man's last victims, his own family. His family—who knew nothing of his crimes until he was caught, and who supported the man through his life—were forced to pick up the pieces after his final act.

5.7 THE MURDER OF JAMES ESSEL

James Essel was born in Ghana and immigrated to the DC Metro area. Mr. Essel and his wife became successful restaurant owners in the District of Columbia and were even featured in local media. After Mr. Essel's restaurant success, he decided he wanted to move to the suburbs in Montgomery County, Maryland, and open another business. Mr. Essel became the proud owner of the Sugarloaf Mountain Market at 23800 Old Hundred Road in Comus.

On Sunday evening, March 22, 1992, Mr. Essel was working at his store in Comus when an unknown assailant entered the store and confronted him. The assailant walked into the store around 5:30 p.m., bought items, and for an unknown reason attacked Essel, savagely stabbing him more than 20 times with a broken wine bottle. Essel appears to have fought back, and an assailant was cut in the process prior to fleeing the scene. Mr. Essel succumbed to his injuries and died on scene. The assailant left DNA evidence behind at the crime scene. This DNA profile was uploaded into CODIS, and no matches were found. Decades of follow-up investigations were conducted in attempts to develop leads that could lead to Mr. Essel's killer. The break in this case came in 2020 when a lead was developed through the utilization of FIGG that assisted in identifying Mr. Essel's murderer. Detectives obtained the suspect's DNA through the authorization of a search warrant. This DNA sample was tested and matched the DNA evidence left behind by the suspect at Mr. Essel's murder scene. An arrest warrant was obtained for the suspect for first-degree murder, felony murder, and robbery with a dangerous weapon. While attempting to make the arrest, a confrontation between detectives and the suspect resulted in detectives using deadly force, and the suspect succumbed to his injuries at the scene.

5.7.1 Case Study Reflections

Mr. Essel's daughter expressed how grateful she was for all the work that investigators had put into her father's case throughout the years. The care and concern the investigators took were remarkable. Decades passed, and at some points, she wondered whether anyone would ever be caught or whether she would ever get answers. Frustration mounted for her, and at times she thought her father's case had been forgotten. After being notified of the identification of her father's murdered she believes now her father can finally rest in peace.

5.8 CONCLUSION

In conclusion, FIGG is one of the most significant tools in major criminal investigations and UHR identification since DNA began to be used by law enforcement. Illustrated in the case summaries presented in this chapter are some of the reasons why FIGG is important. These are only a handful of cases that have been solved by FIGG. Many of these stories outlined heinous acts perpetrated by truly violent offenders. The reflection sections after each case study express how important it is to the victims, their families and friends, and the exonerated innocent when the true offenders in their case have been identified. FIGG provides answers where there were previously questions, closure where there was anguish, and innocence where there was presumed guilt. Used correctly, as an investigative tool, FIGG can solve the previously unsolvable and assist the criminal justice system in arriving at the truth.

5.9 SUMMARY

It is hard as a writer to put into words the feelings that are expressed during the investigations of violent crimes and missing persons cases. Victims and their families often are forced into grieving, having experienced horror, trauma, and fear. Investigators often experience vicarious stress and go through a gamut of emotions of their own while working these cases. Cases that are solved quickly provide answers that assist in the grieving process for most people. The horror and fear often subside into feelings of loss that might never go away, but that may someday be manageable. In cases that grow cold, where leads have gone nowhere, the grief becomes anger, the horror becomes despair, and life as they knew it changes before their eyes. The not-knowing, having no real answers, or the fear that the offender will come back, doesn't allow for much-needed healing to take place.

In the writer's experience as a Cold Case detective, all generations of family members seek answers in open cases of their loved ones. Children, grandchildren, great nieces, and nephews, etc., at times want to know what happened to their beloved relative who was murdered decades earlier. As human beings, we all owe it to these families to do our best to provide the answers to those who sacrificed so much. Investigators should never give up, and citizens should consider providing DNA and OPT-IN to law enforcement cases in genealogy databases. You never know who might be that key genetic lead that can provide a family with that much-deserved closure.

BIBLIOGRAPHY

1. Armstrong, Mia, The Marshall Project (Website), *In an Apparent First, Genetic Genealogy Aids a Wrongful Conviction Case*, written 07/16/2019 updated 07/17/2019. https://www.themarshallproject.org/2019/07/16/in-an-apparent-first-genetic-genealogy-aids-a-wrongful-conviction-case
2. Contreras, Óscar, ABC Denver 7 (Website), *Decades-old Cold Case Solved After Weld County Detectives Identify Remains Found in Field*, written 06/05/2024. https://www.denver7.com/news/front-range/weld-county/decades-old-cold-case-solved-after-weld-county-detectives-identify-remains-found-in-field
3. McRae, Jennifer, CBS News (Website), *Colorado Cold Case Solved 24+ Years After Remains Found in Weld County Field*, written 06/05/2024. https://www.cbsnews.com/colorado/news/cold-case-solved-24-years-colorado-body-found-weld-county-field/

4. Mooar, Brian, The Washington Post, *Woman Found Slain Was Biologist*, written Oct. 05th, 1994.
5. Montgomery County Department of Police, News Release, *Update: Detectives Identify Suspect Involved in Virginia Beach Shooting; Wanted for 1992 Homicide of James Essel*, written February 13th, 2020. https://www2.montgomerycountymd.gov/mcgportalapps/Press_Detail_Pol.aspx?Item_ID=32012
6. Montgomery County Department of Police, News Release, *Cold Case Detectives Arrest Suspect for Series of Rapes; Press Conference to be Held Today*, written September 14th, 2018. https://www2.montgomerycountymd.gov/mcgportalapps/Press_Detail_Pol.aspx?Item_ID=31222
7. Montgomery County Department of Police, News Release, *Cold Case Detectives Identify Suspect of Unsolved Rapes and Murder; DNA and Genealogy Helped Solve the Crimes*, written March 15th, 2019. https://www2.montgomerycountymd.gov/mcgportalapps/Press_Detail_Pol.aspx?Item_ID=31513
8. Montgomery County Department of Police, News Release, *Detectives Make Arrest in Potomac River Rapist Cold Case*, written November 14th, 2019. https://www2.montgomerycountymd.gov/mcgportalapps/Press_Detail_Pol.aspx?Item_ID=31855
9. WTOP, American Nightmare Series, Season Two, Episode 5, *A Break in the Case*. Released March 22nd, 2021. https://americannightmarepodcast.com/episode/s2-ep5-a-break-in-the-case/
10. WTOP, American Nightmare Series, Season Three, (Episodes 1 through 9), *Unknown Subject*, https://americannightmarepodcast.com/shows/season-3/
11. Smugeresky, Steven, SAKI (Sex Assault Kit Initiative), *Genealogy Research: Successful Convictions of Serial Rapist (Webinar)*, May 23rd, 2019. https://www.sakitta.org/webinars/webinar-view.cfm?id=31
12. Correa, Frances, Germantown MD Patch, *Elderly Rape Victims Speak Out*, written February 11th, 2011. https://patch.com/maryland/germantown/amp/981777/elderly-rape-victims-speak-out
13. Morse, Dan, The Washington Post, *Behind the Hunt for a Rapist Who Targeted Older Women: DNA Genealogy and a Cop Named Smugs*, written September 14th, 2018. https://www.washingtonpost.com/local/public-safety/md-detectives-say-theyve-solved-7-year-old-rapes-of-2-elderly-women/2018/09/14/98d0d914-b838-11e8-a2c5-3187f427e253_story.html
14. Hermann, Peter & Morse, Dan, The Washington Post, *Police Arrest Alleged 'Potomac River Rapist' Link to Attacks in Maryland*

and Georgetown, written November 14th, 2019. https://www.washingtonpost.com/local/public-safety/police-arrest-alleged-potomac-river-rapist-in-south-carolina/2019/11/14/2a3b7038-0709-11ea-8292-c46ee8cb3dce_story.html

15. Keegan, Suzie (director), FilmRise (Production Company), *Bloodline Detectives: Rockville Horror*, aired January 28th, 2021.
16. Weld County Sheriff's Office (CO), WCSO 24-year-old Cold Case Solved! John Doe 2000 is Identified, written: June 05, 2024, https://www.weldsheriff.com/Public-Interest/2024-News/WCSO-24-year-old-cold-case-solved-John-Doe-2000-is-identified

CHAPTER 6

FIGG and How It Is Useful in Solving Crimes

Steven Smugeresky

CONTENTS

DOI: 10.4324/9781032714998-6

6.1 INTRODUCTION

In this chapter you will learn how forensic investigative genetic genealogy (FIGG) is generally performed and how it ties into a criminal investigation. You will learn the best practices recommended by the writer, per his experience, and dos and don'ts based on state laws and the DOJ policy on FIGG best practices. You will learn the myths that cause skepticism and methods for developing public relations to ensure your organization is as transparent as privacy rights will allow. You will learn the difference in working unidentified human remains cases as opposed to putative perpetrator cases. You will learn the process from the selection of a case by an investigator to the final resolution and recommendations for future progression of this investigative discipline. The writer will proceed step-by-step from his experience to show how this investigative discipline can be done in the least intrusive way possible while evaluating public safety and public privacy. A case study is provided at the end of the chapter to illustrate how the information in this chapter plugs into a law enforcement investigation. The basic flow of a FIGG investigation is illustrated in Figure 6.1.

6.2 SELECTING A CASE

According to the DOJ guidelines for FIGG best practices, FIGG can be used when: (a) a case involves an unsolved violent crime, and the candidate forensic sample is from a putative perpetrator, or when (b) a case involves what is reasonably believed by investigators to be the unidentified human remains (UHR) of a suspected homicide victim. In addition, the prosecutor may authorize the investigative use of FGGS for violent

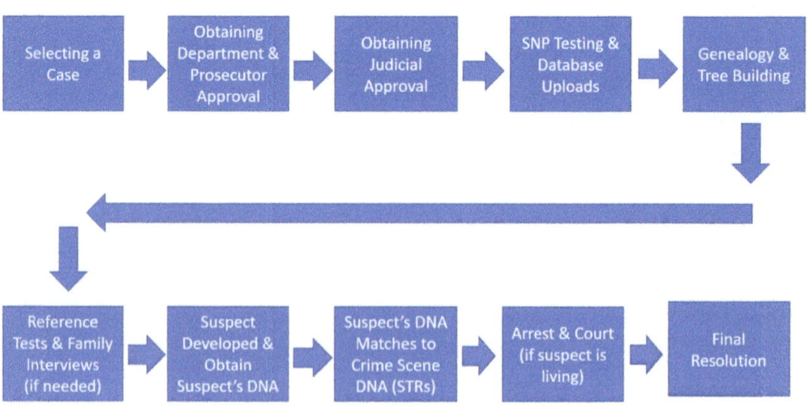

FIGURE 6.1 FIGG workflow from start to finish.

crimes, or attempts to commit violent crimes, other than homicide or sexual offenses (while observing and complying with all requirements of this interim policy) when the circumstances surrounding the criminal act(s) present a substantial and ongoing threat to public safety or national security. According to Maryland state law, FIGG can be used for the following crimes: murder, rape, a felony sexual offense, a criminal act involving circumstances presenting a substantial and ongoing threat to public safety or national security. An exception to Maryland Law exists for UHR and Missing Persons cases, allowing these cases to be initiated without having to pursue judicial approval (took effect Oct. 1, 2024). The GEDMatch Database allows law enforcement to upload SNP DNA raw data for cases involving violent crimes and to identify human remains. The Family Tree DNA data base allows law enforcement to search their data base for the following: "Unidentified Remains & Violent Crime" which means any homicide (including murder or non-negligent manslaughter), any sex crime (including rape), or any of the following serious crimes or criminal offenses: (1) armed robbery, (2) aggravated assault, or (3) any other violent crimes as defined by the FBI Uniform Crime Reporting (UCR) Program.

In the writer's experience, FIGG has been used in murders and both serial rapes/sex assaults and individual rapes/sex assaults, and UHR cases. Start by having your investigative team review a case. This will consist of reading documents and statements, watching any videos, and leafing through photographs. Investigators should locate and evaluate all existing evidence for potential testing and future prosecution. The investigator needs to determine if all leads were exhausted. A simple definition for exhaustion is that each lead has either been disproven or followed out enough that no further information exists to develop or expand the lead. Investigators should have discussions with their crime labs to see if the evidence sample of the putative perpetrator or UHR collected from the crime scene is in CODIS. Once the investigator confirms that the DNA sample is in CODIS, the investigator must confirm that more of the evidence sample exists in order to do future SNP testing.

Check with the lab to make sure that the DNA sample is the sample from the original sample that went into CODIS, or matches the original sample that went into CODIS, to send to the genetic genealogy/SNP testing company that is chosen. Start a FIGG case with the mindset that the investigation takes a finesse approach rather than a "Bull in a China shop" approach. Most individuals whom investigators will interview in a FIGG case are genetic witnesses or family and friends of genetic witnesses. This takes a more personal and understanding approach to an interview. If a person could be closely related to your suspect, an additional layer of questioning should be added to gain as much information as possible, for your suspect's timeline—but you still have to keep in mind that the witness is assisting the investigation and not the involved party.

There are differences in UHR cases and putative perpetrator cases. In UHR cases the investigator is looking for the identity of a deceased person/loved one who may have been a missing person, a homicide victim, or just died and has been unable to be identified. Most times in these cases, investigators can work through family trees and interview many people for possible reference tests without worrying whether or not they're going to tip off a suspect. In putative perpetrator cases, the family trees should be worked as much as they can until the investigator determines it is time to begin reference testing. When selecting a reference tester, the investigator needs to determine how likely it is that this person knows, or actually is, our suspect. These tests are the hardest for investigators because they don't want to knock on a door and step on a "land mine." Investigators need to be careful in choosing their reference testers. Most putative perpetrators in cold cases have put in years of thought and planning for the day an investigator knocks on their door, while the investigator may not be certain that the subject is the perpetrator.

6.2.1 Informed Consent

In regard to FIGG, there are two prongs of informed consent. The first prong of informed consent has to do with the consumer DNA databases that law enforcement uses to upload their SNP evidence profile. In states such as Maryland, the law states that law enforcement may only use the consumer DNA databases that inform their customers that law enforcement *will* search their data base and the customer may *opt-out* of these searches. The second prong of informed consent comes during a reference tester interview. In the state of Maryland, an investigator must inform the reference tester of the FIGG process, the specifics on the testing kit, and where their DNA is going, what happens when their DNA is processed and shared with others, and what type of case the investigator is working, along with some approved case-specific details. One common practice is to utilize the law enforcement agencies' original press release to give to the reference tester to read. It informs the individual of previously approved case information adequately without providing more information than necessary.

6.3 OBTAINING DEPARTMENT APPROVAL

When a case is selected, a detailed review and PowerPoint™/presentation is constructed. This process, while just a suggestion of best practices, is one that has been proven successful in the writer's experience. For states that don't require judicial approval to initiate a FIGG case, an internal selection process like the following example is strongly recommended. This presentation should include a summary of the crime and situation, victimology, theories on the suspect, as well as the leads that have been exhausted and ruled out, and the leads that have been followed and have not been conclusively eliminated. This PowerPoint™ should be presented to your unit of detectives, a lab representative, a ranking investigative supervisor (preferably two supervisors) in your department, and a representative from the prosecutor's office. The ensuing discussion is useful in determining if there is potential prosecutable merit of the case, and, if so, the avenues that should be followed up on at the present time.

6.4 OBTAINING JUDICIAL APPROVAL

Obtaining judicial approval is a process that may vary from state to state. In some states this process is not necessary, and in others, an affidavit, a search warrant, or a court order may be requested to initiate the FIGG case. If a court document is required for judicial approval, the document should include the following:

- A summary of facts of the crime
- The specific crime that qualifies the DNA to be used for FIGG
- The process of exhaustion of leads
- A brief description of how identifying this DNA will further this case.

The process for obtaining your court document will be jurisdiction- or department-specific, but a rough guideline should be the following: (1) an investigative supervisor review and approval of the document, (2) a brief review of the document by the assigned prosecutor, and (3) then a visit to the on-duty judge. Be prepared to explain the process of FIGG to the on-duty judge. FIGG is a relatively new capability with a developing process, and not all judges are familiar with the work being done. Take the time to explain the basic flow of work as previously shown in Figure 6.1.

6.5 SNP TESTING & DATABASE UPLOADS

When selecting a SNP testing company, shop for the work you need done in the case as soon as possible. Get quotes from as many companies as you can and compare their services. Often, rates will vary based on the different services that can be performed. Make sure you understand all the services and testing options that are provided. If the offer is too good to be true, it often is. Ask how their service navigates (i.e., complies with!)

state and federal law. Better yet, ask agencies that have used them in the past to critique their experience and provide their recommendations. The bottom line is that SNP testing and genealogy services can be expensive, but they can be *extremely* expensive if your agency is paying for things it just doesn't need, or they fall into the private sector's "upsell." Make sure the company is only utilizing the consumer DNA databases that practice the informed consent policy and include an opt-out option.

Remember that SNP testing produces the results and raw data that make FIGG effective. Meaning essentially that the raw data and SNP alone won't solve your case. The comparison of raw data to other raw data to find persons who share amounts of DNA with your kit will assist in solving the case.

Conducting a FIGG investigation is not an "IF"; we can bring this to resolution as much as a "WHEN" we will bring this to resolution. This is because, as long as there are people who share DNA in your match list, there are family trees to build and DNA connections to find.

FIGG is not an exact science, but an extremely effective tool to create investigative leads so that the exact science can be applied. Shared persons are obtained from public DNA databases that are open to law enforcement searches, and persons in these databases have opted in to law enforcement searches. Currently, the only two consumer DNA databases that openly inform their consumers and have the OPT-IN and OPT-OUT options are named *GEDMatch* and *Family Tree DNA (FTDNA)*. When a SNP DNA profile is created for your crime scene evidence or your UHR, you may upload the raw data generated from the profile into one or both of the accepted databases for law enforcement (future databases may be limited, closed, created, or allowed as FIGG progresses). When the raw data is uploaded into the consumer database, the investigator or genealogist receives a share DNA "match" list. On this list, the data obtained is limited to how much DNA is shared with the suspect and *not* specific STR profile DNA markers. The amount of DNA shared is measured in a unit called Centimorgan (cM). There are many charts, programs, and websites that you can find on the internet that will provide the amount of centimorgans and their potential position in the family tree. Most charts will show you the traditional layout of a family and the name of the relationship each person has with you. See Figure 6.2, which illustrates a traditional family layout. A shared DNA chart will list within each box for each position/relationship a range of cMs that is equivalent for that position (example 10–250); some charts will list an average as well. See the example of this in the *Parent* box in Figure 6.2.

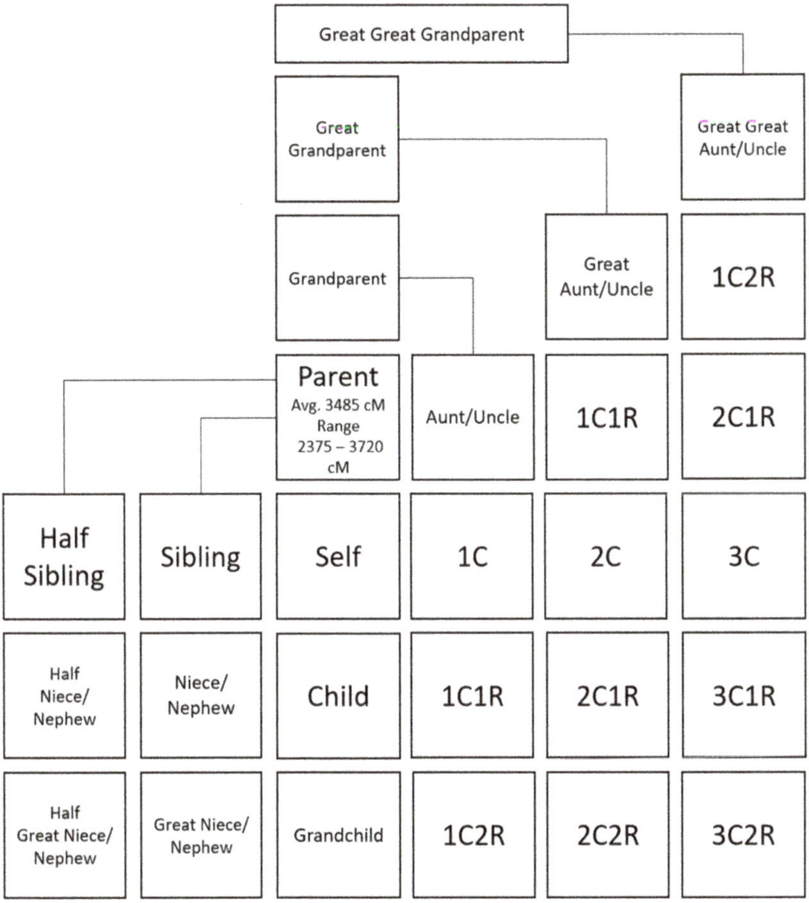

FIGURE 6.2 Family relationship chart.

6.6 GENEALOGY & TREE BUILDING

It is important to understand that genealogy and family tree building are not a newly developed technique. This skill has been around for centuries and has been handed down from generation to generation to the present. In relation to the age of these techniques is how long law enforcement has been using the same skills to make connections in cases. Prior to FIGG, law enforcement utilized genealogy and family research to find witnesses, fugitives, missing persons, connections to suspects—the list goes on and on. The same research skills that are used in criminal investigation also apply to genealogy, and vice versa. Good investigators utilize whatever they have legal access to, and access the information in ways that comply with federal, state, and local laws. This is why it is important to understand the basics of genealogy research and family tree building. In this section, I will briefly go over how to research and build family trees as viewed from an investigator's experience in FIGG. Other chapters in this book will go more in-depth and lay out the concepts and system of how genealogists work on trees. There are so many variables in genealogy, so it's important that an investigator has an open mind when building their family connections.

Family tree building can be as simple as building out to parents, children, grandparents, or first cousins, or as hard as building to great, great, great, grandparents, 4th or 5th cousins, and further. High matches and low matches can also be deceiving. The success in building family trees is based on the documentation and the family connections you find. You can have a high match of a parent or equivalent, but if there is no documentation of who that parent is, you may not find your suspect or UHR easily. It is the same for a lower shared match; if there is a plethora of documentation, and you account for everyone in the family, it may be easy to find your suspect or UHR.

Find a format or template that allows you to understand and explain the information that you gather. Not everyone comprehends, interprets, or understands in the same way as everyone else. Given this, it may take different formats and several times to explain things correctly for others to be on the same page. See Figure 6.3 as an example of building a chart to common ancestors from one shared match. This example shows you the direct connection, from your shared person to your suspect, who may be at a 4th cousin position.

Pick a person from your match list, and see if you can identify them through their email address, a built tree, or by name. Take the amount of cMs shared and plug them into a shared DNA chart, and find all of the family positions and equivalents. Begin building from the shared person to their parents, then grandparents, then great grandparents, and continue until you reach the ancestor you want to start building forward at. Build forward to where your family position is in the cM chart, and assess if this could be the correct generation for your suspect or UHR. If the generation is too young, build your trees on a generation further back; if the persons are too old, build some branches a generation earlier.

Building family trees can be as easy as finding a well-documented family and following a published work of the family that goes back generations, or as difficult as hitting walls on every turn because the family isn't documented, or such documents are missing. The first thing that is needed is that you identify and verify your shared match.

Sometimes you will work a tree and get the feeling, or ultimately prove, that things don't really fit. This is when you need to start thinking about the potential of adoptions, unknown affairs, sperm donations,

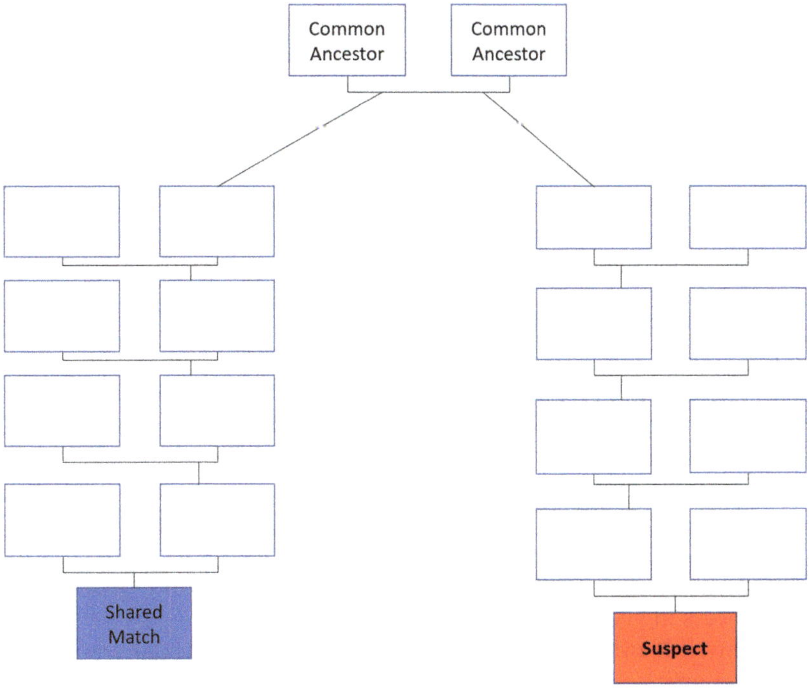

FIGURE 6.3 Examples of building a chart to common ancestors from one shared match.

artificial insemination, and mistakes in the recorded documents. These are just some possible examples of anomalies that may arise in the course of an investigation.

6.6.1 The Writer's Personal Experience Example

"Once I worked a case and began to build a family tree from a very distant shared match. The family was well-documented, and I had built a huge family tree. I worked diligently for four months piecing together chart after chart of family and relatives. As I continued to work, I began to see a pattern emerge with the family origin. The family origin was slightly off from what the DNA recommended. I began to feel something was wrong. I decided to contact the original shared match, to confirm that they were related to person who's DNA match our kit. After a pleasant conversation with this person, they had confirmed my fears that it was not their DNA, but rather it was their child's DNA. Not only that, but their child had been adopted. I was a little discouraged and frustrated, but it wasn't the end of the world. So, I went back to the drawing board and started on new shared match."

A great phrase from a well-respected detective friend of mine that I always come back to in my work is: "If it doesn't fit, it doesn't fit. Don't force it."

Examples of documents that can be utilized to make your family connection can be any of the following records:

Census Records: Census records could be useful to find data on geographical locations, ages, in some cases dates of birth, siblings, parents, grandparents, in-laws, neighbors, occupations, place of birth, and parents' place of birth.

Obituaries: Obituaries and death notices are hidden treasures in genealogy/family history research. Obits can be found in a number of locations: Archived Newspapers, online funeral announcements, funeral home websites, grave locator websites, and library microfilm are just a few. Obituaries can contain information on family relationships, geographical locations where the deceased lived, worked, or activities the deceased may have done. Not all obituaries are the same, so reading each one can unlock key pieces in your family tree.

Archived Newspaper or Periodical Publications: Archived newspaper articles often have achievements and announcements of key family gatherings and events. These articles can help to clarify and confirm family relationships in your trees, along with the information in published family trees, birth and death records, and potential church documents.

Genealogy Web Sites: Genealogy/family history research methods and ideas that can be utilized are the use of genealogy-focused websites for documents and family relationship connections. Examples of these sites are Family Search, Ancestry, Geni, and My Heritage. Family Search is a free website that you can create an account with to utilize and find digitized records from all over the world. Ancestry & My Heritage are genealogy websites with paid versions and library versions that members can access archived document collections and publicly sourced family tree information. Geni is a genealogy website that consists of publicly sourced family trees.

The use of newspaper archives and internet obituary websites is recommended, along with the use of The National Archives and the Library of Congress. Don't be afraid to actually put your boots on the ground and travel to local libraries, funeral homes, county courthouses, cemeteries, and churches or other houses of worship. These places hold a wealth of genealogy information, and much of it is not digitized as of the current day. Social media searches and Google searches can assist in making and clarifying family connections and may also give you hints at where persons have traveled or lived. Once you have an idea where someone has lived, verify records by using regionally specific historical data to confirm your theory. Confirming where your target family comes from can be useful in verifying future information and getting you to the next piece of your tree.

If, for example, you have information that a certain family was from a small town in Poland, you could isolate your searches to this location until you find enough information to widen the search and prove or disprove your data. Get creative with search terms and spellings. Often, names on documents are misspelled or have bad handwriting that is not recognized by search engines, or spellings of names and letters used are confused with other letters and spellings during an indexing process. Of course, it is commonly known that immigrants entering at Ellis Island and other ports of entry frequently had their names truncated or anglicized as well.

So, you may have to be creative with your search terms and name spellings and take the time to leaf through hundreds or thousands of pages of documents to find names that look similar or have similar pronunciations. Knowing what is publicly available and what you can pay for is helpful. Some documents are not allowed to be released by certain government agencies because of privacy issues. In these cases, you have to ask yourself what information you are looking for, and whether it could be found on another easily accessible document through public record or social media.

The bottom line is that genealogy is like building a puzzle; you search for the right piece to fit (some easier than others), and the pieces that don't fit you put on the side until you find the other pieces that connect them. Creativity and persistence are the best skills a genealogist can have.

Often in FIGG cases, genealogy databases and research may only bring you so far. Many of these cases have been solved by using publicly sourced information and traditional genealogy research. On the other hand, many times the key pieces in these cases come down to information that may be included in law enforcement-related databases. This is why it is always important for both the genealogist and the investigator to work together so that key pieces don't get missed. Investigators trained in genealogy research have the best advantage in these cases.

6.7 REFERENCE TESTS & FAMILY INTERVIEWS

There are times in every investigation when investigators hit a wall and need to try something new to develop a lead. This holds true for building family trees in a FIGG case. Most of the time in a FIGG case, two things are done by investigators to develop a family tree beyond what publicly accessible documents can take you. Finding family reference testers and conducting family interviews can often provide information to get you beyond the wall.

6.7.1 Reference Tests

Reference tests are done by obtaining voluntary DNA from members of the potential suspect's or UHR's family to submit and compare to the suspect's or UHR's DNA. When these tests are done correctly, they can be extremely helpful in eliminating or drawing into focus the path to the suspect or UHR. Always inform and educate reference participants of all the information they may need to know in order for them to make a decision on whether they wish to participate in the investigation or not.

6.7.2 Family Interviews

Family interviews are conducted with members of the family on a voluntary basis to obtain more information on family lines and biological relationships, while also obtaining investigative information that may develop a potential suspect or the identity of a UHR. Family interviews can be tricky in putative perpetrator cases. Often, you don't know how close your putative perpetrator may be to your interviewee and what kind of relationship they may have. The interview has to be well-planned, and the interviewer has to be prepared. Informed consent needs to be utilized, and a balance of informative questions, dealing with the case and questions about family connections, has to be developed and phrased in ways that the interviewee feels comfortable and that they feel their information is protected. You never want the family member to feel like a suspect. Keeping the interview nonconfrontational and free-flowing with what the interviewee wants to tell you, while interjecting a few guiding family connection questions, is recommended. The main goal of the interview is to confirm the work that has already been done; if you fill in some of the gaps that were missed, that is "gravy."

When family lines are developed with potential suspects or UHRs, return to the traditional investigative approach. The use of law enforcement and social media databases is essential to develop connections between family, geographical areas, MO of the crime, missing persons, and/or suspects. Law Enforcement only databases include such items as judiciary case search, local police report writing systems, corrections records, arrest reports, and dispatched calls for service. Any of these can help to make critical connections that the private sector genealogist can't because of their lack of access. Social media searches are also important because they can provide the background and narrative to confirm the connections that may not always be clear.

6.8 SUSPECT DEVELOPED & OBTAIN SUSPECT'S DNA

When a suspect is developed in a putative perpetrator case, the investigator must make a decision on how to obtain the suspect's DNA. This decision often involves a discussion with the investigator's team, prosecuting attorneys, crime lab personnel, and undercover surveillance personnel. Two techniques for obtaining the suspect's DNA are obtaining a search warrant and conducting covert operations.

6.8.1 Search Warrant for a Suspect's DNA

When obtaining a Search Warrant for DNA in a FIGG case, work with your command staff and the State's Attorney's Office. You must have building blocks for probable cause. Genealogy is *one* building block. A search warrant should not be written with genealogy as its only basis for probable cause. Other building blocks to utilize are as follows:

- Physical description, and how it compares to the phenotype report or the victim's description of the suspect.
- Geography of where the suspect worked/lived during the time of the crimes, and how it relates to the crime scene or victim.
- Potential for access to victim/victims.
- Pattern of criminal behavior.

6.8.2 Example of Wordage That Can Be Used for a Search Warrant for FIGG

Here is an example of language that could be used to file for a search warrant:

> A genetic genealogy screening identified a subject, who would be the second cousin or genetic equivalent, from the suspect DNA. Genealogical research, law enforcement databases, and family-based interviews were conducted and resulted in the identification of (Place the name of your suspect here) as a second cousin, or genetic equivalent, to the subject listed in the report.

6.8.2.1 Covert Operations

When discussing covert operations with regard to LE (FIGG and non-FIGG related), I believe that it is extremely important to hammer home why covert operations are considered and why they are performed. The #1 reason for any covert operation is to maximize *safety* (safety to the agents/officers, safety of the public, and safety of the suspect that is being sought or apprehended).

Another reason for covert operations is to be the least intrusive to the person or persons identified in an investigation and to rule out potential noninvolved persons and identify involved persons. This is the "less hands in the cookie jar" effect. When more people are involved, and official documents are written and filed, the information has more chances of finding its way to the public via social media, media, and other avenues. Ultimately, as a consequence, that could twist and distort the information, compromise the investigation, and ultimately ruin an innocent person's reputation. To piggyback on the second reason for number two is that covert operations are done to protect privacy; believe it or not, many times a simple covert operation can be conducted quickly and efficiently to produce important information to eliminate the noninvolved parties. This ultimately saves detectives from hours and hours of research, interviews, and surveillance, and the deep dig into someone's life...which is not very private!

6.9 SUSPECT'S DNA MATCHES TO CRIME SCENE DNA (STRS)

When a DNA sample is: (a) obtained from your putative perpetrator, (b) tested and compared to the crime scene DNA, and (c) is a match, this STR comparison is the main probable cause for your arrest warrant. This is comparable to a fingerprint match. Investigators should have their probable cause statement written and reviewed by supervisory staff, as well as their prosecutor in the case, if available. An arrest warrant should be obtained through your jurisdictional process, and a briefing should be set with the apprehension team. The team should be briefed on the case and the suspect's criminal history, MO, access to firearms, work history, vehicles they drive or have access to, anyone the suspect lives with, and any other information that may affect the apprehension of your suspect.

6.10 ARREST & COURT

Once a DNA confirmation test is made and the putative perpetrator is identified, the investigator will obtain charging/arrest documents/warrants for the putative perpetrator. It is at this moment that the case will pick up media attention and recognition. After the media has rendered their thoughts and opinions, a new, long, tough journey begins for the investigator and the prosecutor. The DNA evidence is a piece of the total product when prosecuting these cases successfully. Investigators and prosecutors must track down and re-interview living witnesses and prove, and disprove, given alibis and even investigate known false leads for documented proof. The investigator and prosecutor must provide a lengthy number of items and documents for discovery and be able to counter court motions from the defense. Many of these cases will not see a trial court room for a year-plus.

6.11 FINAL RESOLUTION

The final resolution in such cases can come in many forms, some more satisfying than others. The main hope is for a seamless apprehension of the putative perpetrator, a confession of the incident circumstances, and a guilty plea before the court with a waiver of appeal rights. Often, this best-case scenario does not happen. Because cold cases tend to be on the older side—and may have been perpetrated years ago— sometimes the putative perpetrators identified have already passed on. In other cases, some are still hard-core criminals and decide to go out in a blaze of glory that often results in a shootout, which takes their life and injures others. Still others decide that the weight of a guilty verdict and media attention is too much, and they take their own life. Finally, some are just not in good health from years of hard living, and they die in jail before a trial can be completed. The death of the punitive perpetrator will often lead to resolution, but not always answer the much-wanted questions. Other resolutions from the DNA confirmation could be the existence of an accident, contamination of evidence, an identification of a witness, or a co-conspirator. This can also include the exoneration of a previously convicted defendant.

6.12 RECOMMENDATIONS FOR LEARNING THE SKILLS NEEDED FOR FIGG CASE

One recommendation for FIGG investigators is to take a college-level historical research techniques class. Often, a college-level research class will teach the art of finding non-digitized documents and periodicals that contain valuable information on historical subjects. The use of microfilm, microfiche, and historical directories, reference books, and magazines often yields valuable information for genealogy work.

As stated previously, don't be afraid to get up out of your chair and away from your computer screen. The family information you need can be found in libraries, cemeteries, courthouses, and archive buildings, just to name a few places. Familiarize yourself with how law enforcement investigations work and develop. Learning the steps and legal procedures in a criminal investigation is essential knowledge in FIGG cases. Though the genealogy and family tree building takes the spotlight in most of these cases, it's the investigator who sorts out if the next steps, and any moves to collect DNA samples, and to do it in the legal way for their jurisdiction, so that this evidence and process will hold up in court. Without the knowledge of criminal investigation, a genealogist can become frustrated with the slow, methodical process that law enforcement may take in the collection and testing of DNA and the apprehension of the putative perpetrator. Many times, the genealogist can see the path to the putative perpetrator and explain that all; and all the investigator has to do is collect DNA from this relative. But, in the real world of LE relations its more difficult than just asking (see the section on *Reference Testing*). Finally, learn by doing; never be afraid to ask questions. And, don't be so confident that you rule out other possible options.

6.13 CASE STUDY

The following case will show the practical steps taken in an FIGG investigation and how it relates to the steps and concepts discussed in this chapter. The information in this case study (i.e., the entirety of "Text Box" to follow) is taken in full directly from the New Castle County Police Department's Press release media page. (https://nccpdnews.com/2021/03/23/new-castle-county-police-announce-the-identification-of-a-homicide-victim-from-1977/). Figure 6.4, and its explanation, are additions by the writer created to assist in understanding FIGG in this investigation.

THE MURDER & IDENTIFICATION OF MARIE PETRY HEISER

In June 1977, the body of a female was located in the wooded area off Old Union Church Road (County Road 456) in New Castle County, Delaware, by an area teenager who was traveling home on his bicycle. New Castle County Police, as well as officials from the State of Delaware Medical Examiner's Office, responded to the scene to conduct an investigation. The Delaware Medical Examiner, at the time, determined the manner of the victim's death to be a homicide.

Police conducted an extensive investigation into the woman's murder but were not able to identify her or obtain any leads.

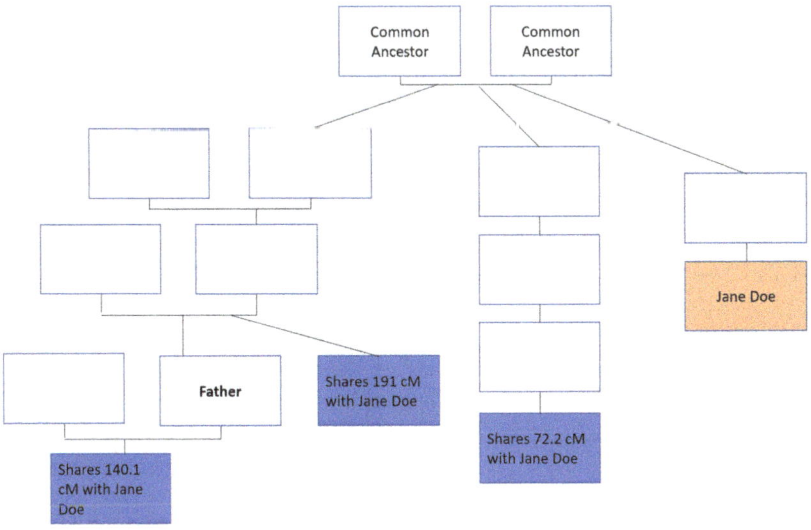

FIGURE 6.4 Shared matches and family reference testers who shared a common ancestor.

The case grew cold. At the time, police relied heavily on fingerprints and dental records in hopes of identifying human remains.

In 2008, information regarding the discovery of the body and surrounding circumstances was reviewed for entry into the National Missing and Unidentified Persons System (NamUs). As part of the case entry process, a DNA profile from the remains was extracted by the University of North Texas for Human Identification and was entered into the Combined DNA Index System (CODIS). There were no matches found in the nationwide system.

In February 2017, New Castle County Police learned of a Virginia company called Parabon Nanolabs, which specializes in DNA phenotyping. Phenotyping is a process of predicting physical traits and ancestry from DNA evidence.

Thom Shaw, IAI Certified Forensic Artist at Parabon Nanolabs, completed a sketch of the victim. The lab subsequently provided a Snapshot Phenotype Report to include a computerized composite of the victim. Parabon Nanolabs further submitted the genetic information to ancestry databases and provided genetic genealogy analysis, which included a potential family tree for the victim.

New Castle County Police report that the breakthrough in this investigation occurred when Montgomery County, Maryland,

Officer Steven Smugeresky became involved in the case. Members of the Cold Case Squad consulted with Officer Smugeresky after he conducted a presentation at a law enforcement Cold Case Seminar. In 2019, Officer Smugeresky took over the ancestry research of the victim's DNA profile. He worked tirelessly on potential family trees in an attempt to develop information on the identity of the remains. New Castle County Police followed up on these leads and obtained DNA samples from possible relatives. Figure 6.4 illustrates the lead that was generated from the relatives that chose to reference test.

The Delaware Division of Forensic Science DNA Laboratory expeditiously conducted analysis of the possible relative samples and extracted DNA profiles. The profiles were entered into CODIS, and a positive comparison to the 1977 remains was received.

Based on that analysis, the identity of Marie Petry Heiser was confirmed by Chief Medical Examiner Dr. Gary Collins of the Delaware Division of Forensic Science. Dr. Collins concurred with the initial Pathologist, ruling Heiser's manner of death to be homicide.

Marie Petry Heiser is from the 6200 block of Crafton Street in Philadelphia, Pennsylvania. It was determined that Marie Petry Heiser was 50 years old at the time of her death and had not been reported missing to any police agency.

Heiser was the wife of William Heiser Sr., who was a member of the Philadelphia Police Department's Highway Patrol from the 1950s until the early 1960s. They were known for performing thrill shows in area stadiums and arenas. Heiser Sr. had to leave the Police Force after being seriously injured while rehearsing for one of these shows. He then worked as a truck driver for an area transportation company. William Heiser Sr. relocated to the area of South Daytona Beach, Florida, in the late 1970s and subsequently died in 2006.

Marie Petry Heiser was a homemaker and part-time employee at the former Ashbourne Country Club in Cheltenham, Pennsylvania, in the 1970s. She was the mother of two children. The children had been told by their father that their mother had packed her bags and left the residence in Philadelphia, leaving behind no information on where she had gone.

The New Castle County Police are continuing their investigation into Marie Petry Heiser's death. We are specifically interested in anyone who may have known the Heiser family, especially from the area of Crafton Street in Philadelphia or the former Ashbourne Country Club in Cheltenham, Pennsylvania, as well as the greater South Daytona Beach area in Florida.

Anyone who has any information about this incident, please contact Detective Jeffrey Sendek at 302-395-8110 or via email at Jeffery.Sendek@newcastlede.gov or Lt Teresa Williams (RET) at 302-395-8110 or Teresa.Williams@newcastlede.gov or by calling the New Castle County Division of Police non-emergency number at 302-573-2800.

Citizens can text a tip anonymously simply by texting the number 847411 keyword: NCCDE. You can also submit a tip via our website at http://www.nccpd.com. Tipsters may also call Crime Stoppers at (800) TIP-3333 or visit us on Facebook (IM) at New Castle County Division of Police.

6.14 CONCLUSION

In conclusion, FIGG is just another investigative lead. FIGG wields a huge amount of responsibility in balancing the issues of public safety and public privacy in order to bring a violent criminal to justice. A FIGG case must always be done with the most informed consent provided, and obtained, as possible. Before taking on a FIGG case, make sure you have a good understanding of your jurisdiction's laws and the DOJ guidelines for best practices, in addition to meeting with your prosecutors and command staff. Without their approval and understanding of the process, cases can stall and never make it to the finish line. UHR cases and putative perpetrator cases should be worked similarly in the process, but differently when it comes to reference testing. Reference testing in UHR cases can be done more frequently, and more family questions can be asked without worrying about tipping off a suspect.

6.15 SUMMARY

Since the identification of the Golden State Killer in 2018, the use of FIGG has been propelled to the front line of Cold Case and UHR investigations, currently being used in hundreds of cases. In this short time, the development of FIGG has led to suggested best practices, state laws, and guidelines to utilize this investigative discipline within the confines of current laws. This chapter has outlined the process and best practices of a FIGG case from start to finish based on the writer's experience, Maryland state law, and the DOJ guidelines pertaining to FIGG. This chapter explains the considerations for selecting a case, the legal process involved in beginning a case, the importance of reference testing and informed consent, and the considerations for search warrants or covert operations. A case study from the New Castle County Police Department has been provided that highlights how FIGG fits into the process of a criminal investigation.

BIBLIOGRAPHY

1. Bettinger, Blaine, The Genetic Genealogist Adding DNA to the Genealogist's Toolbox. Updated 2020. https://thegeneticgenealogist.com/wp-content/uploads/2020/03/Shared-cM-Project-Relationship-Chart.png
2. United States Department of Justice, United States Department of Justice Interim Policy Forensic Genetic Genealogical DNA Analysis and Searching. https://www.justice.gov/olp/page/file/1204386/dl#:~:text=A%20suspect%20shall%20not%20be,profile%20previously%20uploaded%20to%20CODIS
3. Maryland Governor's Office of Crime Prevention and Policy, Forensic Genetic Genealogical DNA Analysis and Search (FGGS). https://gocpp.maryland.gov/crime-statistics/law-enforcement-reports/fggs/
4. New Castle County Police Department (DE), Police Announce the Identification of a Homicide Victim from 1977, March 23, 2021. https://nccpdnews.com/2021/03/23/new-castle-county-police-announce-the-identification-of-a-homicide-victim-from-1977/
5. GEDMatch, GEDmatch & Community Safety. https://www.gedmatch.com/community-safety/
6. FamilyTreeDNA, FamilyTreeDNA Law Enforcement Guide Last modified: April 11, 2024, Effective Date: April 25, 2024. https://www.familytreedna.com/legal/law-enforcement-guide
7. Library of Congress, Chronicling America. https://www.loc.gov/collections/chronicling-america/about-this-collection/
8. Guerrero, Juan Carlos. ABC7 News, *TIMELINE: A look back at the Golden State Killer's crime spree that lasted a decade*, https://abc7news.com/golden-state-killer-joseph-deangelo-confesses-how-was-the-discovered-who-did-kill/3391867/

CHAPTER 7

Notable Cases Cleared Through Forensic Investigative Genetic Genealogy (FIGG)

Tracey Leigh Dowdeswell

CONTENTS

DOI: 10.4324/9781032714998-7

7.1 INTRODUCTION

Since its inception, forensic investigative genetic genealogy (FIGG) has cleared many high-profile and notorious criminal cases. As of July 2024, public data indicates that FIGG has been used to clear over 1400 cases, including about 400 cases in which a crime scene sample was used to identify the suspect in one or more criminal offenses, as well as the identification of nearly 700 decedents [1, 2]. There are likely other cases that have not been made public. There are, therefore, many notable cases cleared through FIGG—all of which present unique features and opportunities for learning.

Here, I have chosen to focus on cases that are prominent for what they can teach investigators and practitioners about the development of the field, particularly the ideas and practices that are pushing its limits, expanding its capabilities, and breaking new ground in forensic investigations.

I first review some of the oldest human remains identifications and criminal cases that have been cleared through FIGG. I then discuss cases that have been cleared using challenging DNA samples, such as highly degraded and very low copy number DNA. Some of these cases obtained usable SNP profiles from remains that had been in the water for a lengthy period of time or had been subject to burning. Next, I review some innovations in bioinformatics, including direct SNP-to-SNP comparison and forensic phenotyping, and how these techniques have been used to clear cases. This is followed by a discussion of innovations in traditional genealogical research that have broken new ground, as well as the use of FIGG to address several miscarriages of justice.

7.2 CLEARING THE COLDEST OF CASES

Many very old cases have been cleared by FIGG, including the identification of several individuals who have been deceased for more than a century. This is not surprising—as discussed below in Section 7.4.1, the laboratory techniques used in FIGG were pioneered in the field of ancient DNA. If a SNP profile suitable for genetic genealogy can be developed, and there are sufficient familial matches and genealogical records available, then there are very few limitations on clearing even the very coldest of cold cases.

7.2.1 Human Remains Identifications

There have been several human remains identifications that have been cleared through FIGG where the decedents passed away in the nineteenth century. There have been several cases of unidentified children who have been cleared from the 1950s, including one case in which the identification of the young boy also resulted in the identification of his killers and the clearance of that open criminal case.

7.2.1.1 The Identification of Richard Bunce

The earliest case cleared to date is the identification by the DNA Doe Project of a set of pre-Civil War era human remains that were discovered in Hudson, Ohio, in August of 2019, when a pine coffin was unearthed during construction work for a real estate development [3]. Inside the box was a set of skeletonized human remains that were old enough to have become partially fossilized [3].

The remains have been identified as Richard Bunce (also spelled Bunts and Buntz), who was born in 1783 [4]. He died in Hudson in 1852 and was buried in a small cemetery on the family property that contained several graves from the 1850s [3].

Given the age and condition of the remains, developing a SNP profile was challenging [4]. Astrea Forensics performed several DNA extractions from Richard's femur and humerus [4]. They were able to extract DNA in the range of about 107 to 412 nanograms—although this represents all of the DNA extracted, including contaminant DNA, and not only human DNA [4].

The DNA was then sequenced by HudsonAlpha Discovery [4]. The mapping rates showed that about 3–4% of the extracted DNA belonged to Richard, meaning that the human DNA in the samples was in the range of about 3.85 to 18.65 picograms [4]. The final sequencing results covered less than 15% of the human genome [4]. Kevin Lord of Saber Investigations then used several bioinformatics techniques to make the most of these results, including mixture assessment and imputation-based variant calling [4]. The final mapping rate of human DNA was about 40% [4].

One of the key matches used to solve this case was Richard's third great-granddaughter [3]. She shared approximately 169.1 centimorgans (cM) of DNA with Richard, and she still lived in the Hudson, Ohio area [4]. The shared cM Project indicates high probabilities that this amount of shared DNA is in the range of a second cousin once removed, or a half great-great aunt or uncle relationship [5].

A combination of genealogical records, land survey maps from the 1850s, and kinship testing with living relatives enabled volunteers from the DNA Doe Project to solve this case in a little over two years [3].

7.2.1.2 The Identification of Esther Ann Granger

A set of human remains from the 1860s that has been identified through FIGG is that of Esther Ann Granger—a 17-year-old girl whose skull was hidden in the walls of James Skinner's home in Batavia, Illinois [6]. When Skinner removed a baseboard in November of 1978 for some home renovation work, her skull tumbled out from the wall [7]. Esther's skull was then sent for forensic examination to the Anthropology department at Northern Illinois University and from there to the Batavia Depot Museum [7]. It was forgotten until March of 2021, when museum officials rediscovered it while auditing their collection [7].

In early 2023, the skull was sent to Othram by the Kane County Coroner's Office and the Batavia Police Department [7]. Othram then extracted a DNA profile, and its in-house genetic genealogists identified Granger [7]. She was born on October 26, 1848 [7]. She had been married at the age of 16 to her husband, Charles Granger [7]. She died in 1866 from complications due to childbirth and was buried in Merrillville, Indiana [7]. She may have been the victim of grave robbing for anatomical specimens [6]. This was not uncommon at the time, as physicians would often pay for specimens to study [6].

Othram made the final confirmation of Granger's identity through kinship testing with her second great-grandson, 69-year-old Wayne Svilar, who is himself a retired cold case detective in Portland, Oregon [6]. She has now been laid to rest in Batavia, with her descendant, Svilar, delivering her eulogy [6].

7.2.1.3 The Identification of Chester Alfred Breiney

The oldest human remains identification that has resulted in the identification of the victim's killer was cleared by Othram in November of 2024 [8]. 7-year-old Chester Alfred Breiney was identified as the child whose remains were discovered on October 4, 1959, in a culvert in Mequon, Wisconsin [8]. The autopsy found that Chester had been subjected to significant neglect and physical abuse before he died, including a healed broken rib [8].

Chester Breiney was born to Josephine Breiney on February 26, 1952 [8]. He was surrendered to the Good Will Farm orphanage and then adopted in March of 1955 by William and Hilja Jutila, who renamed him Markku Jutila [8]. The Jutilas later moved from Houghton County to Chicago—but without their adopted son [9]. Relatives then reported the boy missing [9].

During the investigation, the Jutilas were arrested and confessed to police that they had disposed of Chester's body in a ditch in Mequon on their way to Chicago [8]. Hilja told police that she had beaten the boy to death [9]. In November of 1966, prosecutors dropped the charges against them for lack of evidence, as they could not identify the body recovered in Mequon as that of Breiney [9]. The Jutilas passed away in 1988 [9].

A funeral was held for Breiney on November 15, 2024, at the St. Peter of Alcantara Church [8]. He was laid to his final rest at the St. Mary's Parish Cemetery in Port Washington, Wisconsin [8].

7.2.1.4 The Identification of Joseph Augustus Zarelli

Breiney's case follows upon several prominent cases that identified missing and murdered children from the 1950s. One of these is the identification of the "Boy in the Box," discovered on February 25, 1957, in a wooded area of northeast Philadelphia [10]. He had been beaten to death and then wrapped in a blanket and placed inside a JCPenny bassinet box [10].

Misty Gillis and Colleen Fitzpatrick led a team that included Identifinders and the Vidocq Society; they identified him as 4-year-old Joseph Augustus Zarelli [10]. Like Breiney, Joseph was adopted, perhaps informally, and those responsible for his murder have not yet been identified [10].

7.2.1.5 The Identification of David & Derek D'Alton

The same holds true for the identification of the "Babes in the Woods"—discovered on January 15, 1953, in Vancouver's Stanley Park—as 6-year-old David D'Alton and his 7-year-old brother Derek D'Alton

[11]. A groundskeeper was clearing brush in the park near Beaver Lake when he found the skeletal remains of the two boys under a woman's fur coat [11]. They had been bludgeoned to death with a hatchet, which was recovered at the scene. Investigators believe the boys were killed in the park sometime in 1947 or 1948 [11].

The brothers were identified by Redgrave Research Forensic Services, using a SNP profile developed by the Paleo-DNA Laboratory at Lakehead University in Thunder Bay, Ontario [11]. The boy's mother is described as a very loving woman who likely had nothing to do with the boy's death [11]. Like Breiney and Zarelli, the D'Alton brothers may have been taken into care away from their birth families, but there is currently little information on their past or on the identity of their killer [11]. Their identification will hopefully move the investigation forward.

Sometime in the 1990s, the boys' remains were taken out of the Vancouver Police Museum by Detective Brian Honeybourn, now retired from the Vancouver Police Department's homicide unit, because he felt that their display was disrespectful [12]. He took them to Dr. David Sweet at the Bureau of Legal Dentistry at the University of British Columbia, and a tooth was extracted from each of the skulls [12]. The remains, minus the skulls, were then cremated and interred in the ocean with the assistance of the police chaplain [12].

The SNP profiles that gave the boys back their identity were developed from these teeth—a fitting result from a gesture of kindness and respect [12]. This also demonstrates the extent to which many of the older remains identified through FIGG have been misplaced, mistreated, and mishandled throughout their long journeys.

7.2.2 Criminal Identifications

Criminal FIGG investigations typically use a DNA profile developed from a suspect's DNA that has been collected from a crime scene. Therefore, the criminal cases that have been cleared using FIGG do not go back as far in time as human remains identifications. Nonetheless, there have been seven criminal cases cleared from the 1960s, and many dozens of cases cleared from the 1970s [1]. The earliest criminal case resolutions go back to the 1950s [1].

7.2.2.1 Kenneth Gould: The Murder of Duane Bogle & Patty Kalitzke

The earliest criminal case cleared to date through FIGG is the identification of Kenneth Gould as the lead suspect in the January 2, 1956, murder of Duane Bogle, 18, and Patty Kalitzke, 16, in Great Falls, Montana [13].

The couple had been together in what is now Wadsworth Park, and was then known as a "lovers lane" [13]. Their remains were found the following day. Bogle's body lay face down beside his vehicle; his hands had been tied behind his back, and he had been shot in the head [14]. Kalitzke had been taken several miles away, where she was sexually assaulted and then shot [14].

Samples were collected from Kalitzke's body during her autopsy [13]. One of those samples was a slide prepared from a vaginal swab that contained a small amount of sperm, and that had fortunately been preserved through the years [13]. An STR DNA profile was produced by the Montana State Crime Lab and entered into CODIS in 2001 [13]. About 36 suspects had their DNA samples taken and compared with the DNA profile, all of whom were ruled out—including serial killer Edward Wayne Edwards and notorious mafia enforcer "Whitey" Bulger [13].

A SNP profile was developed in 2019 by Bode Technology, and FIGG was used to identify Gould as the source of the sperm [13]. Gould passed away in 2007, and his remains were cremated [14]. The STR profile developed from the autopsy slide was then compared with Gould's surviving children; kinship testing confirmed that they were likely the children of the man who left the sperm that was collected from Kalitzke's remains [14]. Gould had lived near where her body was found, and he sold the family property and moved away shortly after the murders [14].

7.2.2.2 John Reigh Hoff: The Murder of Candy Rogers

Another prominent case from the 1950s that was cleared through FIGG is the identification of John Reigh Hoff as the lead suspect in the March 6, 1959, rape and murder of 9-year-old Candice ("Candy") Rogers in Spokane, Washington [15]. This was Spokane's oldest cold case before being solved by Othram in 2021 [15].

Candy had been a Camp Fire Girl, and she was selling mints door-to-door when she went missing [15]. Hundreds of people from the community joined in to search for her [15]. This produced its own tragedy when three members of the Air Force who had joined in the search died when their helicopter hit a power line and crashed [15].

Candy's body was found about two weeks later in a remote area in the brush, hidden under branches and pine needles [15]. She had been raped and strangled with a piece of her clothing [15].

Semen was recovered from Candy's clothing, and a DNA profile of the suspect was developed in 2000 and entered into CODIS, but there were no hits, and the case went cold once again [15]. In March of 2021, Othram produced a SNP profile from the semen and used FIGG to narrow down the suspect list to just three brothers, all of whom were deceased, and only one of whom was survived by a child [15].

Police then obtained a DNA sample from John Hoff's daughter, Cathie, and they confirmed through kinship testing that she was the daughter of the man who was the source of the semen left on Candy's body [16]. Hoff's body was then exhumed for DNA testing; a direct comparison found that it was 25 quintillion times more likely that he was the source of the semen than an unrelated individual drawn at random from the population [17].

Hoff would have been 19 years old at the time of Candy's murder. He was convicted of a similar sexual assault in 1961, in which he had bound the victim and strangled her, and for which he served six months in prison [17]. He died by suicide in 1970 [16].

7.3 INNOVATIONS IN SNP PROFILING

Since its inception, FIGG has pushed the boundaries in terms of developing SNP profiles from degraded DNA. This is one of the chief advantages of using SNP profiles over more traditional DNA profiles, as it can be difficult to generate usable STR DNA profiles from samples when the DNA is degraded or available only in low amounts [18]. Laboratories working with FIGG have managed to clear several cases using very low copy numbers and highly degraded DNA.

Here, I have chosen to highlight some notable cases where the remains had been especially poorly preserved, but SNP profiles were nevertheless able to be obtained, and the individual was identified. In some of these cases, the remains had been in the water for a great length of time or had been burned—both of which will result in significant degradation of DNA [19].

7.3.1 The Identification of Darren Roy Marchand

The identification of Richard Bunce, discussed above in Section 7.2.1.1, shows the promise of SNP profiling followed by genetic genealogy as a means of identifying human remains—even when there are only picograms of human DNA available for sequencing [4].

In July of 2021, Othram identified Darren Roy Marchand as the lead suspect in the murder of 14-year-old Stephanie Anne Isaacson [20]. Isaacson was sexually assaulted and strangled to death on June 1, 1989, in Las Vegas, Nevada, on her way to school that morning [20]. Her father phoned the school, worried when she did not come home; her remains were later found in a vacant lot that was on her usual route to school [20]

Semen was collected from Isaacson's shirt; STR DNA profiles were developed in 1998 and again in 2007, and entered into CODIS [21]. By the time the sample came to Othram in November of 2020, there were only about 15 individual sperm cells remaining—for a total of about 150 picograms of human DNA [20, 21].

This is one of the lowest recorded amounts of DNA known to have successfully cleared a criminal case [20]. The optimal amount of DNA to amplify STR markers is about 1 nanogram, given a maximum of 28–30 PCR cycles; forensic STR profiles cannot be obtained using traditional PCR-based methods when there are low DNA copy numbers [18].

Marchand committed suicide six years after the murder, at the age of 29 [21]. Before the rape and murder of Isaacson, Marchand had been charged with the similar rape and murder of 24-year-old Nanette Vandenburg [21]. Those charges were dismissed due to a lack of evidence [21]. The DNA found on Vandenburg's body was recently compared with that of Marchand, and this case has also been cleared with Marchand as the lead suspect [21].

Had DNA evidence been available at the time, he may have been convicted of Vandenburg's murder, and the killing of Isaacson could have been prevented. This serves as a reminder of the power of DNA evidence and the role it can play in preventing similar miscarriages of justice.

7.3.2 The Identification of Ada Elizabeth Fritz

In August of 2023, Moxxy Forensics completed the challenging identification of 61-year-old Ada Elizabeth Fritz, whose remains were found on May 18, 1976, in Grand Bay, Alabama [22]. She was discovered floating in Sessions Creek by two teenage boys who had gone there to fish [22]. She had been shot in the back of her head, and her hands had been severed [22].

After she could not be identified, her remains were donated to science and then cremated [22]. This treatment of unidentified remains was not uncommon for the time [1]. Intermountain Forensics was able to develop a SNP profile from a dental mold taken at Fritz's autopsy, using the M-Vac® system—like her hands, her dentures had also been removed by her killer, likely to prevent her identification [22].

The M-Vac® System is a wet vacuum system that can be advantageous in collecting challenging DNA samples, such as touch and low copy number DNA, as well as DNA deposited on fabrics and porous surfaces. [23]. This method collected 145 nanograms of DNA from Fritz's dental mold [22]. However, the DNA was degraded, and so whole-genome sequencing was used to develop a usable SNP profile [22].

The results of the genealogical analysis were confirmed through kinship testing with Fritz's nephew—her only living relative [24]. Police suspect that Fritz was killed by Henderson James Williams, who died in

2010 while serving a prison sentence for killing his mother [22]. Williams had cut off his mother's hands and burned them; they were recovered from a smoldering fire in his backyard [22].

7.3.3 The Identification of Rodney Peter Johnson

In August of 2020, Othram identified Rodney Peter Johnson, whose remains were found by a fisherman on June 11, 2020, floating in Lake Stickney in Snohomish County, Washington [25]. He had been murdered, having been killed by a gunshot wound to his head [25]. He had likely been in the lake for several years, weighted down by some heavy material; this allowed for the development of adipocere, and which posed a significant challenge for extracting a usable DNA profile [25].

Adipocere is caused by the post-mortem decomposition of adipose tissue into a variety of fatty acids [26]. This is most likely to take place in anaerobic environments, such as underwater [26]. David Mittleman of Othram stated that they were working with one-fifth of a nanogram of DNA—or the equivalent of about 20 cells—with about 98% of the DNA in the sample consisting of bacterial contamination [25].

Johnson's mixed ancestry also made it difficult to complete an accurate forensic facial reconstruction [25]. He was originally thought to be partly African American, with perhaps an admixture of Asian ancestry [25]. Several forensic sketches were made, depicting him variously as a younger and an older African American man, and as a stout Asian man [25]. Forensic phenotyping showed that Johnson is mainly of European biogeographical ancestry, but also has some Native American ancestry, being descended from the Lummi tribe of the Pacific Northwest through his Great-Great-Grandfather Thomas She-Kla-Malt [25].

7.3.4 The Identification of Burned Remains: Frank Little & Zachary Wells

Burned and charred remains also present significant challenges to extracting and sequencing DNA and completing a successful identification through FIGG. Despite this, there are several cases that have successfully identified burned remains. This section will focus on two particularly challenging identifications.

The first is the identification of Frank ("Frankie") Little Jr. by the DNA Doe Project in 2021 [27]. Workers at a business on Cannon Road in Twinsburg, Ohio, discovered his skull behind the business; a garbage bag containing further skeletonized remains was located nearby [27]. He had likely been deceased for several years [27]. Little had been killed by stabbing and blunt force trauma; his remains were then dismembered and set on fire [27]. This post-mortem treatment helped to obscure his identity, and he became known for the next 40 years only as Twinsburg John Doe [27].

The case also presented special challenges since Little was African American, and there were few matches on the GEDmatch and Family-TreeDNA databases for the genetic genealogists at DNA Doe to work with [27]. Traditional genealogy is challenging in African American communities, as their enslaved ancestors were often not named in slave schedules or estate documents, making it very difficult to document family histories prior to emancipation in 1865 [28].

The team at DNA Doe discovered that his family had a connection to Laurens County, South Carolina [27]. In the end, volunteers constructed over 70 family trees to organize over 4,000 distant relative matches [27]. Finally, they were able to confirm his identity through kinship testing with his brother, who was then living in Georgia [27].

Little had been a guitarist and songwriter for the popular 1960s R&B band, *The O'Jays*, who were inducted into the Rock and Roll Hall of Fame in 2005 [29]. His former bandmates report that he gave up a promising music career to move back to Cleveland to be with the woman he loved [29]. His daughter has passed away, but he is survived by a son. Police continue to investigate his murder [29].

In 2022, Othram successfully developed a DNA profile and identified Zachary Wells, 59, who had been found deceased in January of 2022 after a house fire on Oak Drive South West in Atlanta, Georgia [30]. Reports indicate that the home was abandoned, and Wells had been taking shelter there in unsafe conditions [30]. This case also demonstrates the utility of SNP profiling to develop usable genetic information even from badly burned remains.

7.4 INNOVATIONS IN BIOINFORMATICS

There have been several innovations in bioinformatics that have advanced the field. This section will highlight some notable cases that were cleared using direct SNP-to-SNP comparisons and the use of forensic phenotyping to advance investigations.

7.4.1 SNP-to-SNP Comparison

DNA samples prepared from rootless hair have been used to clear several cases to date, including criminal and human remains identifications [1]. The conviction of David Allen Dalrymple in 2024 for the 1982 murder of 9-year-old Daralyn Johnson in Nampa, Idaho [31] also broke new ground in that it used a direct SNP-to-SNP comparison, rather than a traditional forensic STR comparison, to calculate the final likelihood ratio [32]. In this case, a rootless hair was the only usable forensic evidence that had been preserved after Charles Fain was wrongfully convicted of the offense, and the case was closed in 1984 [33].

A mitochondrial DNA sample taken from the hair in 2001 was used to exonerate Fain [33]. Reports indicate that an STR DNA profile could not be obtained from the hair, but a usable SNP profile was produced [34]. Therefore, this case could not be cleared using existing law enforcement databases; Dalrymple had been convicted of a sexual offense in 2004, and so his DNA was most likely already in CODIS [35].

This identification of Dalrymple was accomplished through a partnership between Barbara Rae-Venter and Dr. Ed Green of the University of California at Santa Cruz [32]. Green's team has recently published a paper describing the IBDGem method of directly comparing two SNP profiles and calculating a likelihood ratio on their degree of relatedness—including the likelihood that the two profiles come from the same person [36].

Dalrymple's defense team moved to exclude the DNA evidence on the grounds that the forensic application of these techniques was novel, and Green's lab at UC Irvine was not accredited [32]. This case demonstrates the close connection between the techniques of "ancient DNA" and FIGG: Green trained in the laboratory of Svante Pääbo at the Max Planck Institute for Evolutionary Anthropology, where many of these techniques were pioneered, including the use of "ultra-clean rooms" for the prevention of contamination [32]. Pääbo was awarded the Nobel Prize for his work on ancient DNA in 2022 [37].

In June of 2024, Justice Whitney of the Canyon County District Court ruled that Dr. Green's testimony was admissible, setting a significant precedent for the admissibility of this evidence in future cases [32]. He found that the science behind the extraction of SNP profiles from rootless hair, as well as the statistics behind the direct SNP-to-SNP comparison in this case, was sufficiently reliable to be put to the trier of fact [32].

7.4.2 Forensic Phenotyping & Biogeographical Ancestry Determination

IBDGem is not the only notable innovation in bioinformatics that has played a role in clearing cases. Once a SNP profile has been obtained, forensic phenotyping can be used alongside genetic genealogy to assist in case resolution by enabling investigators to determine some of the externally visible characteristics of the subject—such as hair, skin, and eye color, as well as facial features [38]. DNA profiles may also be used to determine the biogeographical ancestry of the subject; this, in turn, can assist in identification by enabling the creation of a more accurate facial reconstruction, and can also advance genealogy research [39].

One such case is that of Perrean Gray, who was identified by Parabon Nanolabs in 2019 [40]. Her charred remains were found in a dumpster in Sacramento in June of 2001 [40]. The fire was so hot that the metal dumpster had partially melted, and since soot was located in her lungs, she was likely alive when the fire was started [1].

A facial reconstruction was performed at the time of her death, and a clay bust was produced and released to the public [40]. However, this early reconstruction had incorrectly depicted her as a light-skinned Caucasian woman and so was noticeably different from her true appearance [40].

In April of 2023, Othram identified a set of remains that were found in May of 2020 in Galveston, Texas, as 24-year-old Calvin Mbwambo [41]. His remains were recovered from Galveston Beach near Kahala Drive, where he had most likely drowned [41].

Mbwambo was originally thought to have been African American, but a bioinformatics analysis of his DNA showed that he was of recent East African ancestry, mainly Bantu, and with likely ancestral ties to Kenya or Somalia [41]. This enabled investigators to identify him as Mbwambo—an exchange student from Tanzania whose family had reported him missing [41].

One of the most well-known identifications through FIGG is Barbara Rae-Venter's identification of Joseph James DeAngelo, known as the "Golden State Killer," as being responsible for a series of rapes and murders in California [42]. Information that the suspect had Italian ancestry was important to Rae-Venter's investigation, as was information obtained from GEDmatch that the suspect likely had blue eyes [42].

Phenotypic and ancestry determinations can focus the investigation on certain family lines, narrow down candidates for identification, test, and eliminate certain hypotheses. In this way, they can be very useful in advancing FIGG investigations.

7.5 INNOVATIONS IN GENEALOGICAL RESEARCH

FIGG depends for its efficacy not only on advances in laboratory techniques and bioinformatics, but also on highly skilled traditional methods of genealogical research. This section will focus on two cases that were particularly innovative: the identification of David Sinopoli as the lead suspect in the murder of Lindy Sue Biechler, and the identification of C.L. Daniel—the first victim to be identified from the mass graves associated with the 1921 Tulsa Race Riot.

7.5.1 David Sinopoli: The Murder of Lindy Sue Biechler

David Sinopoli was arrested in July of 2022 for the rape and murder of 19-year-old Lindy Sue Biechler, which took place in her apartment in Manor Township, Pennsylvania, on December 5, 1975 [43]. Biechler had been sexually assaulted and stabbed to death by an intruder, using one of her own kitchen knives [43]. She had been stabbed multiple times and had received numerous defensive wounds in the struggle; the knife was still located in her neck [44].

The genetic genealogical work in this case was conducted by CeCe Moore, the Chief Genetic Genealogist at Parabon Nanolabs [43]. A DNA profile of the suspect was developed from semen collected from Biechler's underwear in 1997 [44], as well as blood collected from her pantyhose [43].

In most genetic genealogical investigations, relative matches will be used to identify a most recent common ancestor (MRCA), and descendancy research will then identify the descendants of the MRCA; the person of interest will most likely be found among the descendants [45].

In this case, however, no MRCA could be identified. Moore found that only very distant relative matches were being returned, with under 30 cM of shared DNA [44], and therefore, the MRCA ancestral couples likely went back to the 1600s or 1700s [43]. Moore could not solve this case in the usual way, and she had to find a new approach [43].

She discovered that many of the subjects' ancestors migrated from the small village of Gasperina—in the southern Italian region of Calabria—to Lancaster, Pennsylvania [43]. After months of searching, Moore found the records of a local club whose residents came from Italy, and whose membership cards listed where members were born [43]. Moore searched through the records to look for those who immigrated to Lancaster from Gasperina [43].

Moore then cross-referenced the membership cards with other genealogical records, such as immigration documents from Ellis Island, and draft registration cards from the two World Wars, in order to identify the living descendants of each of the Gasperina immigrants who were living in Lancaster in 1975 [43]. She identified Sinopoli as the most likely subject [43].

Police then retrieved cast-off DNA from a coffee cup that Sinopoli had discarded and made the arrest [43]. Sinopoli, then 69, pled guilty to third-degree murder in October of 2023 and was sentenced to 25 to 50 years in prison—nearly 47 years after the crime took place [46]. Moore states that she has used this method in subsequent cases. This case demonstrates how innovative genealogical research can overcome the hurdles presented by even very distant relative matches.

7.5.2 The Identification of C.L. Daniels, Deceased in the 1921 Tulsa Race Riot

In July of 2024, the team at Intermountain Forensics successfully identified C.L. Daniel as a victim of the 1921 Race Riot in Tulsa, Oklahoma [47], and who was buried in a mass grave along with more than 100 others [47]. About 300 African American residents of a prosperous community in Tulsa were killed, and more than 1200 homes, churches, and businesses were burned in a series of racially motivated riots that lasted for days [47].

The effort to excavate and identify the unknown remains began in 2020 [47]. DNA profiles from about 30 sets of human remains have been developed from remains recovered from a mass grave of about 120 individuals linked to the massacre [47]. Genealogist Alison Wilde of Intermountain Forensics located a letter in the National Archives from Daniel's mother in which her attorney seeks veterans' benefits from his service in World War I [47]. That letter stated that Daniel was killed in the Tulsa Race Riot [47]. The final confirmation of his identity came from kinship testing with the descendants of his brothers [47]. Genetic genealogy, combined with traditional methods of genealogical documentation, has a great facility to piece together incidents like this—even after more than a century has passed [47].

While we have seen the identification of unidentified remains, disaster victims, and even living individuals, this is the first successful identification of a victim in a mass grave. Human remains identifications using CODIS algorithms depend upon a very close relative—usually in the first degree—already being in the database [48].

In the US, CODIS searches produce a hit to a profile already in the database only about 30–50% of the time [48]. Moreover, even when identifications can be made using CODIS, there is a false negative rate, meaning that the algorithm fails to correctly detect a close relationship when present [48].

Based on the original 13 CODIS loci, the false negative rate for parent-child relationships is about 14%, and this can go above 42% for full-sibling pairs [48]. FIGG has the potential to overcome these limitations and successfully resolve these identifications, even after a long period of time has passed or when close relatives are not available.

7.6 REMEDYING MISCARRIAGES OF JUSTICE

FIGG has been used to remedy several miscarriages of justice. This section will address the use of FIGG to remedy wrongful convictions, including the recent exoneration of Robert and David Bintz, as well as the identification of Johnny Lee Coy as the lead suspect in the rape and murder of Rashell Ward—a case that could not be cleared due to the return of a false negative result when the DNA profile developed from the crime scene was entered into CODIS.

7.6.1 Wrongful Convictions & Exonerations

To date, there have been 13 criminal cases cleared through FIGG in which an innocent individual had previously been wrongfully convicted of the offense [1]. There are many more cases in which an innocent person was long suspected of an offense, and which blighted their lives.

For example, Mark Douglas Burns (known as the "Clearfield rapist") was identified through FIGG and pled guilty to a series of rapes in Utah and Wyoming, as well as to the murder of Sue Ellen Gunderson Higgins, who was killed in her home in Evanston, Wyoming, in July of 2001 [49]. Her husband, Sean Higgins, was prosecuted for her murder. Although the charges were dropped before he was convicted, he spent many years of his life and over $70,000 on his defense, at a time when he was coping with the loss of his wife and raising their young child [49].

To date, four individuals have been exonerated through FIGG [1], including the exoneration of brothers Robert and David Bintz by the Investigative Genetic Genealogy (IGG) Center at Ramapo College [50]. Chris Tapp spent 20 years in prison for the 1996 murder of Angie Dodge in Idaho Falls, Idaho, before CeCe Moore of Parabon Nanolabs used FIGG to identify Brian Dripps, who pled guilty to Dodge's murder in June of 2021 [1].

Ricky Davis was exonerated after serving nearly 15 years for the 1985 murder of Jane Hylton in Eldorado Hills, California [1]. Davis was finally freed after Michael Green was identified through FIGG and arrested in February of 2020 [1]. Green pled guilty to Hylton's murder in October 2022 [1].

Robert and David Bintz were convicted in 2000 for the August 1987 murder of 44-year-old Sandra Lison [51]. Lison went missing after finishing her shift as a bartender at the Good Times tavern in Green Bay, Wisconsin [51]. On August 4, 1987, horseback riders found her remains along a riding trail in the Machickanee Forest [51]. She had been sexually assaulted, bludgeoned, and killed by manual strangulation [52].

The Bintz brothers had been drinking at the Good Times tavern the evening she died [51]. Blood and semen found on Lison's remains excluded the brothers [52]. However, prosecutors dismissed the probative value of this evidence at trial, changing their theory of the case to deny that Lison had been sexually assaulted [52]. Instead, prosecutors argued that the Bintz brothers killed Lison because they had a dispute with her over their beer tab [51]. The brothers were convicted by a jury and sentenced to life in prison [51].

The Wisconsin Innocence Project reopened the case and, in 2006, produced a DNA profile of the blood found on Lison's body [51]. This confirmed that the blood came from the same individual as the semen and therefore supported the theory that she had been violently sexually assaulted [51]. This new evidence did not exonerate the brothers because, again, it did not disrupt the prosecution's theory that the sexual activity was not connected with Lison's murder [51].

The Great North Innocence Project took up the case in 2019 [50]. Bode Technologies produced a SNP profile, and in the summer of 2023, the Ramapo IGG Center took over the genetic genealogy research [50]. It took only a matter of days for staff and students at an IGG bootcamp to identify William Joseph Hendricks as the most likely source of the biological evidence left on Lison's remains [50].

Hendricks had been convicted of a similar sexual assault and had only been out of prison for seven months before Lison was killed [50]. He passed away in a psychiatric hospital in 2000—the same year that the Bintz brothers were convicted—and so the case has been cleared by exception [50]. The final confirmation of Hendrick's identity was made through exhuming his remains and comparing his STR profile directly with that recovered from Lison's remains [50].

Several cases cleared through FIGG involve wrongful confessions by Henry Lee Lucas [1]. For example, Lucas wrongfully confessed to the October 4, 1978, rape and murder of 16-year-old Krisann Baxter in Spokane, Washington [53]. Othram identified Keith Lindblom as the likely source of the biological evidence collected from Baxter's remains at her autopsy [53].

Lindblom had been charged in 1975 with a similar sexual assault of a 16-year-old girl in the area, but the charges were dropped to a simple assault [53]. Lindblom was released from prison on August 7, 1978, only a few months before he killed Baxter [53].

7.6.2 A False Negative Result in CODIS

Many cases cleared by FIGG show, like Lindblom, how the criminal justice system's poor handling of cases involving sexual violence allows sex offenders to progress in the severity of their criminal behavior. When one individual is wrongfully suspected or convicted of a criminal offense, the true perpetrator remains at large, giving them the opportunity to harm others [54].

In October of 2023, the FBI's IGG Task Force identified Johnny Lee Coy as the lead suspect in the rape and murder of 14-year-old Rashell

Ward in Red Bluff, California [55]. Ward was abducted around 8 in the morning on March 3, 1983, as she was on her way to school [55]. Her remains were found later that morning [55]. She had been bound, raped, and then shot [55].

Astrea Forensics developed a profile of the suspect from a rootless hair found on Ward's remains [55]. This included a forensic STR profile, which was uploaded to CODIS databases [55]. This profile returned no hits—despite the fact that Coy had been convicted for the 1989 abduction and sexual assault of a mother and daughter, and his DNA profile was in CODIS [55]. Coy was sentenced to life in prison for this crime, and he passed away in 2019 [55].

As discussed above, samples run through CODIS can produce false negative results, where they fail to match a profile that is in fact in the database [48]. Coy's sample was taken in the early days of CODIS when it used only 13 STR loci—on January 1, 2017, this was increased to 20 core loci—and so may not have registered as a hit to a contemporary sample [55, 56].

After the hit, a new STR profile of Coy was produced and entered into CODIS, and this new profile produced a hit to the hair sample collected from Ward's remains [55].

Coy's identification can be compared with that of Dalrymple, discussed above. Public reports indicate that an STR profile of high-enough quality to be entered into CODIS databases could not be obtained from the available evidence [34]. In the case of Coy, an STR sample was uploaded to CODIS, but produced a false negative, most likely because the original sample was developed using older technology [55].

As of the end of 2023, 84 serial offenders had been identified through FIGG [1]. Based on their known victimizations, had these offenders been identified and apprehended after their first known offense, then they could have been prevented from killing 53 people and sexually assaulting 241 victims [1]. The cost of these crimes to society, using McCollister et al.'s estimates, is about $505 million in 2008, or about $737 million in 2024 [57]. Much of these costs represent intangible costs, including the pain and suffering of the victims, and the loss of their ability to contribute to society [57].

Budowle et al. have recently conducted a cost-benefit analysis of using FIGG, including next-generation SNP sequencing [58]. They found that the use of these technologies could prevent up to 50,000 crimes in the US each year, including preventing an estimated 258 people from being murdered, and 8,700 people from being sexually victimized [58]. They found that the annual cost savings could amount to US$4.8 billion per year and justify a greater investment in these technologies [58].

7.7 SUMMARY

To date, FIGG has cleared several high-profile and notorious criminal cases and has made several highly challenging human remains identifications. This chapter has reviewed a selection of cases that have much to teach us about how the field has advanced, and will continue to develop, to clear some of the most challenging cases, including cases involving highly degraded DNA, or low copy number samples, and human remains that have been subject to burning.

Innovations in bioinformatics and genealogical research have also pushed the boundaries of the field. David Dalrymple was convicted of murder in Idaho using a SNP profile developed from a rootless hair that had previously exonerated Charles Fain, who had been wrongfully convicted of this crime. For the first time, this case used direct SNP-to-SNP comparison, and this expert evidence was ruled admissible during the evidentiary hearings in that case. This is a significant development in the field, and one that will surely generate further research into the statistical methods used, and the admissibility of this evidence will be ruled upon by further courts in the years to come.

FIGG has also been used to help remedy several miscarriages of justice, and it has been instrumental in achieving four exonerations to date. When miscarriages of justice like this take place, the true perpetrators may remain at large, giving them the opportunity to victimize others. In identifying dangerous individuals early in their criminal careers, some of this victimization can be prevented. This represents a significant cost savings for society—in addition to the important goals of preventing victimization and bringing missing persons home to their families.

REFERENCES

1. Dowdeswell TL. Forensic genetic genealogy project v. JULY2024. *Mendeley Data, V1 [Internet]*. 2024 Sep 14 [cited 2025 Jan 15]. Available from: https://doi.org/10.17632/z5khb77xyd.1
2. Dowdeswell TL. Forensic genetic genealogy: A profile of cases solved. *Forensic Science International: Genetics*. 2022 May 1;58:102679.
3. DNA Doe Project. Hudson John Doe 2019 [Internet]. March 10, 2020 [cited 2025 Jan 13]. Available from: https://dnadoeproject.org/case/hudson-john-doe-2019/
4. Lord K. Case study: Identifying a pre-civil war john doe. *CrimeBench [Internet]*. 2022 Jan [cited 2025 Jan 13]. Available from: https://crimebench.com/case-study-identifying-a-pre-civil-war-john-doe/

5. Bettinger BT. The shared cM project 4.0 tool v4 [Internet]. c2025 [cited 2025 Jan 13]. Available from: https://dnapainter.com/tools/sharedcmv4
6. Martin S. Cold case solved: DNA helps identify skull found in wall as Indiana teen who died in 1866. *USA Today [Internet]*. 2024 Oct 26 [cited 2025 Jan 13]. Available from: https://www.usatoday.com/story/news/nation/2024/10/26/esther-granger-indiana-teenager-skull-dna/75838400007/
7. Vogen M. Kane county coroner's office and batavia police department team with Othram to identify 1978 Jane Doe. *DNASolves.com [Internet]*. 2024 Oct 24 [cited 2025 Jan 13]. Available from: https://dnasolves.com/articles/esther-granger-illinois/
8. Vogen M. Ozaukee county sheriff's office & Wisconsin DOJ team with Othram to identify a 1959 murdered child. *DNASolves.com [Internet]*. 2024 Nov 8 [cited 2025 Jan 13]. Available from: https://dnasolves.com/articles/chester-alfred-breiney-wisconsin/
9. Thompson C, Reinhart M. Cold case cracked: DNA evidence links Michigan boy, 7, to skeletal remains found in 1959. *Detroit News [Internet]*. 2024 Nov 14 [cited 2025 Jan 13]. Available from: https://www.detroitnews.com/story/news/local/michigan/2024/11/14/dna-evidence-links-michigan-boy-7-to-skeletal-remains-found-in-1959/76274108007/
10. Shapiro E. 'Boy in the Box': Victim finally id'd in Philadelphia's oldest unsolved homicide case. *ABC News [Internet]*. 2022 Dec 8 [cited 2025 Jan 13]. Available from: https://abcnews.go.com/US/boy-box-victim-finally-idd-philadelphias-oldest-homicide/story?id=94687156
11. Lindsay B. For nearly 70 years, these two murdered boys were known as the 'Babes in the Woods,' now they finally have names. *CBC News [Internet]*. 2022 Feb 15 [cited 2025 Jan 13]. Available from: https://www.cbc.ca/news/canada/british-columbia/babes-in-the-woods-identified-1.6352438
12. Kergin B. Babes in the woods: Vancouver police release identities, details about historic murders. *Vancouver is Awesome [Internet]*. 2022 Feb 15 [cited 2025 Jan 13]. Available from: https://www.vancouverisawesome.com/local-news/babes-in-the-woods-vancouver-police-release-identities-details-about-historic-murders-5067386
13. CBS/Associated Press. Investigators solve 1956 killings of Montana teens. It may be the oldest case cracked using DNA and forensic genealogy. *CBS News [Internet]*. 2021 Jun 10 [cited 2025 Jan 13]. Available from: https://www.cbsnews.com/news/montana-1956-killings-patricia-kalitzke-duane-bogle-solved-dna/?fbclid=IwAR2cHpZfVXH0ru1iPqoCpY4erwwAhBV9Y40vB-0C58vsYdEzAWzTlKpplY8

14. McGonigal T. Cold case: 1956 murder of two people in Great Falls finally solved. *KRTV 3 News [Internet]*. 2021 Jun 8 [cited 2025 Jan 13]. Available from: https://www.krtv.com/news/cold-case-1956-murder-of-two-people-in-great-falls-finally-solved
15. Vogen M. Spokane police partner with Othram to identify the 1959 murder of Candice Rogers. *DNASolves.com [Internet]*. 2021 Nov 19 [cited 2025 Jan 13]. Available from: https://dnasolves.com/articles/candice_rogers/
16. Sarnoff M. Spokane police solve 1959 cold case of 9-year-old girl who was raped and murdered while selling mints. *Law & Crime [Internet]*. 2021 Nov 19 [cited 2025 Jan 13]. Available from: https://lawandcrime.com/crime/spokane-police-solve-1959-cold-case-of-9-year-old-girl-who-was-raped-and-murdered-while-selling-mints/
17. Briggs N. Candy rogers cold case murder solved. *Spokane City [Internet]*. 2021 Nov 19 [cited 2025 Jan 13]. Available from: https://my.spokanecity.org/police/news/2021/11/19/candy-rogers-cold-case-murder-solved/
18. Alaeddini R, Walsh SJ, Abbas A. Forensic implications of genetic analyses from degraded DNA—A review. *Forensic Science International. Genetics* 2010;4(3):148–157. Available from: https://doi.org/10.1016/j.fsigen.2009.09.007
19. Mckinnon M, Henneberg M, Higgins D. A review of the current understanding of burned bone as a source of DNA for human identification. *Science & Justice* 2021 Jul 1;61(4):332–338.
20. DNA Solves.com. Las Vegas metropolitan police department and Othram partner to identify 1989 murderer of Stephanie Isaacson." *DNASolves.com [Internet]*. 2021 Jul 21 [cited 2025 Jan 13]. Available from: https://dnasolves.com/articles/stephanie_isaacson_murder/
21. Torres-Cortez R. DNA testing reveals killer in 1989 strangling of Las Vegas Schoolgirl. *Las Vegas Sun [Internet]*. 2021 Jul 21 [cited 2025 Jan 13]. Available from: https://lasvegassun.com/news/2021/jul/21/dna-testing-reveals-killer-in-1989-strangling-of-l/
22. McCarter O. Identifying Jane Doe X-1. *The ISHI Report [Internet]*. 2023 Nov [cited 2025 Jan 13]. Available from: https://promega.foleon.com/theishireport/the-ishi-report-november-2023/jane-doe-x-1
23. Vickar T, Bache K, Daniel B, Frascione N. The use of the M-Vac® wet-vacuum system as a method for DNA recovery. *Science & Justice* 2018;58(4):282–286. Available from: https://doi.org/10.1016/j.scijus.2018.01.003
24. Moxxy Forensics. Case solve: Mobile county Jane Doe identified as Ada Elizabeth Fritz. *Moxxy Forensics [Internet]*. c2023 [cited 2025 Jan 13]. Available from: https://www.moxxyforensics.com/solves

25. Hutton C. Another ID through DNA: Who killed Rodney Johnson? *Herald.net [Internet]*. 2020 Aug 13 [cited 2025 Jan 13]. Available from: https://www.heraldnet.com/news/another-id-through-dna-what-happened-to-rodney-johnson/
26. Magni PA, Lawn J, Guareschi EE. A practical review of adipocere: Key findings, case studies and operational considerations from crime scene to autopsy, *Journal of Forensic and Legal Medicine* (2021). Available from: https://doi.org/10.1016/j.jflm.2020.102109.
27. DNA Doe Project. Twinsburg John Doe 1982. *DNA Doe Project [Internet]*. 2019 Jun 7 [cited 2025 Jan 13]. Available from: https://dnadoeproject.org/case/twinsburg-john-doe-1982/
28. David LT. Supporting the use of genetic genealogy in restoring family narratives following the transatlantic slave trade. *American Anthropologist* 2023 Oct 25;126(1):153–157. Available from: https://doi.org/10.1111/aman.13939.
29. Jones B. Remains found in 1982 identified as guitarist for R&B group The O'Jays. *News 5 Cleveland [Internet]*. 2021 Dec 14 [cited 2025 Jan 13]. Available from: https://www.news5cleveland.com/news/local-news/oh-summit/remains-found-in-1982-identified-as-guitarist-for-r-b-group-the-ojays
30. Vogen M. Fulton county medical examiner's office teams with Othram to identify an Atlanta home fire victim. *DNASolves.com [Internet]*. 2022 May 27 [cited 2025 Jan 13]. Available from: https://dnasolves.com/articles/Fulton_County/
31. Brizee A. Idaho man, in prison when he was charged, found guilty in Nampa girl's 1982 murder, rape. *Idaho Statesman [Internet]*. 2024 Jun 26 [cited 2025 Jan 13]. Available from: https://www.yahoo.com/news/idaho-man-prison-charged-found-002922781.html?guccounter=1&guce_referrer=aHR0cHM6Ly93d3cuZ29vZ2xlLmNvbS8&guce_referrer_sig=AQAAALH2Qn3c0TZgnFArqMc_1FA7IEN6u5DMDKjYnAsJ
32. State of Idaho v. David Allen Dalrymple. Case CR14-20-7840. Court's Ruling on Admissibility of Dr. Green's Testimony, held on June 18, 2024, Before the Honorable Thomas W. Whitney District Court Judge. Third Judicial District of the State of Idaho, Canyon County.
33. Gross A. Charles Fain. *National Registry of Exonerations [Internet]*. 2012 Jun [cited 2025 Jan 13]. Available from: https://www.law.umich.edu/special/exoneration/Pages/casedetail.aspx?caseid=3209
34. ISHI. Debunking myths surrounding law enforcement's use of forensic genetic genealogy. *ISHI Blog [Internet]*. 2022 Jan 12 [cited 2025 Jan 13]. Available from: https://www.ishinews.com/debunking-myths-surrounding-law-enforcements-use-of-forensic-genetic-genealogy/

35. Terhune K, Lewis T. Man charged in 1982 murder of Tampa 9-year-old. *KTVB7 News [Internet]*. 2022 Jan 13 [cited 2025 Jan 13]. Available from: https://www.ktvb.com/article/news/crime/daralyn-johnson-david-dalrymple-murder-1982/277-8fb25df6-366e-4017-be6b-79f6e9882186
36. Nguyen R, Kapp JD, Sacco S, Myers SP, Green RE. A computational approach for positive genetic identification and relatedness detection from low-coverage shotgun sequencing data. *Journal of Heredity*. 2023 Sep 1;114(5):504–512. Available from: https://doi.org/10.1093/jhered/esad041.
37. Curry A. Ancient DNA pioneer Svante Pääbo wins Nobel Prize in Physiology or Medicine. *Science Insider [Internet]*. 2022 Oct 3 [cited 2025 Jan 13]. Available from: https://www.science.org/content/article/nobel-prize-physiology-or-medicine-2022
38. Dabas P, Jain S, Khajuria H, Nayak BP. Forensic DNA phenotyping: Inferring phenotypic traits from crime scene DNA. *Journal of Forensic and Legal Medicine*. 2022 May 1;88:102351. Available from: https://doi.org/10.1016/j.jflm.2022.102351.
39. Schneider PM, Prainsack B, Kayser M. The use of forensic DNA phenotyping in predicting appearance and biogeographic ancestry. *Deutsches Ärzteblatt International*. 2019 Dec;116(51–52):873. Available from: https://doi.org/10.3238/arztebl.2019.0873.
40. Taraya J. Body burned in dumpster fire 18 years ago has been identified through DNA. *Fox News 40 [Internet]*. 2019 Nov 26 [cited 2025 Jan 13]. Available from: https://fox40.com/news/local-news/body-burned-in-dumpster-18-years-ago-has-been-identified-through-dna/
41. Vogen M. Galveston Medical Examiner's Office partners with Othram to identify 2020 John Doe found on the beach. *DNASolves.com [Internet]*. 2023 Apr 14 [cited 2025 Jan 13]. Available from: https://dnasolves.com/articles/galveston-beach-remains/
42. Rae-Venter B. *I know who you are: How an amateur DNA sleuth unmasked the Golden State Killer and changed crime fighting forever*. New York: Ballantyne Books; 2024.
43. Shapiro E. Suspect arrested in 1975 murder after genetic genealogist turns to new approach. *ABC News [Internet]*. 2022 Jul 19 [cited 2025 Jan 13]. Available from: https://abcnews.go.com/US/suspect-arrested-1975-murder-genetic-genealogist-turns-approach/story?id=87061566
44. Taylor M. DNA from coffee cup confirms genetic genealogy suspect in 1975 murder of teen. *Forensic Magazine [Internet]*. 2022 Jul 20 [cited 2025 Jan 13]. Available from: https://www.forensicmag.com/588288-DNA-from-Coffee-Cup-Confirms-Genetic-Genealogy-Suspect-in-1975-Murder-of-Teen/

45. Greytak EM, Moore C, Armentrout SL, Genetic genealogy for cold case and active investigations. *Forensic Science International.* 299 (2019). Available from https://doi.org/10.1016/j.forsciint.2019.03.039.
46. David ND Sinopoli pleads guilty to murdering Lindy Sue Biechler; 1975 death was Lancaster County's oldest cold case. *Lancaster Online [Internet].* 2023 Oct 19 [cited 2025 Jan 13]. Available from: https://lancasteronline.com/news/local/david-sinopoli-pleads-guilty-to-murdering-lindy-sue-biechler-1975-death-was-lancaster-countys-oldest/article_68ba4448-6dc0-11ee-8c9c-77574d6dd705.html
47. Miller K. A World War I veteran is first Tulsa Race Massacre victim identified from mass graves. *AP [Internet].* 2024 Jul 12 [cited 2025 Jan 13]. Available from: https://apnews.com/article/tulsa-1921-race-massacre-graves-identified-c7cdcd5ca4e54ec27bbb836beae86404
48. Ge J, Budowle B. Forensic investigation approaches of searching relatives in DNA databases. *Journal of Forensic Sciences* 2021;66:430–443. Available from: https://doi.org/10.1111/1556-4029.14615.
49. Herbets A. 'Clearfield rapist' avoids death penalty in new murder case, receives life in prison. *Fox 13 News [Internet].* 2020 Dec 4 [cited 2025 Jan 13]. Available from: https://www.fox13now.com/news/local-news/clearfield-rapist-avoids-death-penalty-in-new-murder-case-receives-life-in-prison
50. Gurney D, Mayer JR. Genetic genealogy can stop violent criminals and free the wrongly convicted. *Scientific American [Internet].* 2024 Nov 21 [cited 2025 Jan 13]. Available from: https://www.scientificamerican.com/article/genetic-genealogy-can-stop-violent-criminals-and-free-the-wrongly-convicted/
51. Mathew B. Two Green Bay brothers exonerated in 1987 homicide; DNA match leads to new suspect. *Green Bay Press Gazette [Internet].* 2024 Sep 26 [cited 2025 Jan 13]. Available from: https://www.greenbaypressgazette.com/story/news/crime/2024/09/26/how-a-dna-match-exonerated-bintz-brothers-in-killing-of-sandra-lison/75398361007/
52. Possley, M. David Bintz. *National Registry of Exonerations [Internet].* 2024 Oct 11 [cited 2025 Jan 13]. Available from: https://www.law.umich.edu/special/exoneration/Pages/casedetail.aspx?caseid=6883
53. Teneyck A. Major crimes detectives use genetic genealogy to solve 45-year-old cold case in Spokane County. *KHQ NBC News [Internet].* 2023 Sep 8 [cited 2025 Jan 13]. Available from: https://www.khq.com/news/major-crimes-detectives-use-genetic-genealogy-to-solve-45-year-old-cold-case-in-spokane/article_1db0cdbe-4e8f-11ee-9ae4-4355375ac978.html

54. Weintraub JN, Bernstein KM. Identifying and charging true perpetrators in cases of wrongful convictions. *The Wrongful Conviction Law Review* 2020;1:181.
55. Rootless TM, Hair sample, genealogy help ID suspect in 1983 murder of 14-year-old. *Forensic Magazine [Internet]*. 2023 Oct 25 [cited 2025 Jan 13]. Available from: https://www.forensicmag.com/608130-Rootless-Hair-Sample-Genealogy-Help-ID-Suspect-in-1983-Murder-of-14-Year-Old/
56. Wilson L. Combined DNA Index System: CODIS. In Houk MM, ed. *Encyclopedia of forensic sciences*, 3d ed. Elsevier, 2023. p. 596–604. Available from: https://doi.org/10.1016/B978-0-12-823677-2.00209-9
57. McCollister KE, French MT, Fang H. The cost of crime to society: New crime-specific estimates for policy and program evaluation. *Drug and Alcohol Dependence*. 2010 Apr 1;108(1–2):98–109. https://doi.org/10.1016/j.drugalcdep.2009.12.002
58. Budowle B, Arnette A, Sajantila A. A cost–benefit analysis for use of large SNP panels and high throughput typing for forensic investigative genetic genealogy. *International Journal of Legal Medicine*. 2023 Sep;137(5):1595–1614. Available from: https://doi.org/10.1007/s00414-023-03029-7

CHAPTER 8

Professional Ethics for Practitioners of Forensic Investigative Genetic Genealogy (FIGG)

Tracey Leigh Dowdeswell

CONTENTS

 DOI: 10.4324/9781032714998-8

8.1 INTRODUCTION

Forensic Investigative Genetic Genealogy (FIGG) combines within it all the ethical issues involved in forensic laboratory science, the use of forensic DNA, the ethics of consumer genomics and ancestry determination, the genetic analysis of traits, as well as the possible discovery of unsettling information that may accompany any genealogical research. FIGG poses new ethical issues, even in addition to these. An exploration of the various ethical issues that arise from the forensic use of genetic data can (and does) fill up several volumes.

Therefore, I have chosen in this chapter to focus on emerging norms as embodied in professional codes of ethics, practice standards, and policies that are being developed to guide the ethical practitioner of FIGG. I will also draw upon some recent developments in case law and regulations that shed light on the ethical practice of FIGG, although a full exploration of legal developments in this area is outside the scope of the present chapter.

I have organized the discussion of professional ethics within a broader framework of the key values that underlie the professional standards: individual autonomy, privacy, human dignity, and the need for high standards of professional integrity, competence, and quality of the genealogical work product. These fundamental values and goals serve as a compass to guide the ethical practitioner of FIGG and help to resolve the ethical dilemmas that may arise in the practice of FIGG.

8.2 ETHICS CODES & PRACTICE STANDARDS

At present, there is a growing number of professional bodies that issue licenses and develop professional standards for genetic genealogists. The body most specifically dedicated to the forensic application of genetic genealogy is the Investigative Genetic Genealogy Accreditation Board (IGGAB), which was established to provide a professional credential for practitioners of FIGG, to hold practitioners to high standards of ethics and proficiency, and to provide for a system of accountability for any breaches of the professional standards [1].

In 2025, the IGGAB introduced a proficiency examination of candidates and successful candidates may receive the Accredited Investigative Genetic Genealogist (AIGG) credential [1]. After a public consultation process, the IGGAB issued its current Code of Professional Ethics (IGGAB Ethics Code) [2] and their Professional Standards and Accreditation Requirements (IGGAB Professional Standards) [3] which guide the conduct of AIGGs.

The IGGAB Code of Ethics and Professional Standards are built upon the earlier Genetic Genealogy Standards developed by the Genetic Genealogy Standards Committee, which were published in January of 2015 [4]. Those standards require genetic genealogists to maintain competency, for example, with respect to the different tests available, and to abide by applicable terms of service (including privacy policies) of the direct-to-consumer DNA testing companies that offer genetic genealogy services (Genetic Genealogy Databases) [4]. The Genetic Genealogy Standards also set norms surrounding informed consent and respect for privacy [4].

In 2019, the Board for Certification of Genealogists (BCG) issued the second edition of their Genealogy Standards (BCG Standards), which were revised in 2021 [5]. The BCG Standards set out the Genealogists' Code of Ethics (BCG Code of Ethics) [6], as well as several standards specific to genealogists' use of DNA [5]. The BCG Standards and Code of Ethics are incorporated by reference into the IGGAB Professional Standards [3]. Whether or not they are certified by the BCG, most genealogists will have studied and will be familiar with the BCG Standards as part of their training in genealogical research.

The BCG has recently introduced the Certified Genetic Genealogist^{SM} credential [7] for those who wish to undertake professional genetic genealogical research on behalf of others, and who must have already achieved the credential of a Certified Genealogist® with the BCG [7]. There are many licensing bodies in the field of genealogy, but the BCG is the first to issue a professional board-certification specifically in genetic genealogy [7]. This credential is directed toward serving private clients, rather than the forensic use of genetic genealogy, and so does not include standards or ethical duties specific to FIGG.

The Council for the Advancement of Forensic Genealogy (CAFG) certifies forensic genealogists with the Forensic Genealogy Credential[SM] [8]. The CAFG was established to educate, mentor, and provide professional credentials to forensic genealogists who produce research reports for cases with legal implications, and who may be called upon to testify in court [8]. While the CAFG does not focus on genetic genealogy or on FIGG, it has issued a Code of Standards of Practice and Conduct [8]. The CAFG Standards require forensic genealogists, for example, to avoid recruiting heirs or other beneficiaries, to refuse contingency-fee arrangements, to avoid other conflicts of interest, to attain a high level of truth and accuracy in their forensic research, to comply with law, and to enter into written service agreements for all work undertaken [8]. These are common and sound requirements for many professions.

One forensic organization has put forward substantive guidelines governing ethical conduct in FIGG programs. The National Technology Validation and Implementation Collaborative (NTVIC) published a set of guidelines for establishing FIGG programs in 2023 [9]. The NTVIC then received feedback from stakeholders and published a set of updated guidelines (NTVIC Guidelines) [10]. The NTVIC Guidelines establish criteria for the ethical practice of FIGG, such as defining which cases qualify for investigation, the necessity of practitioners to abide by the terms of service of Genetic Genealogy Databases and the declared privacy preferences of their users, as well as the protection of privacy and confidentiality, and the need for practitioners to obtain informed consent from Reference Test Takers [10].

The NTVIC Guidelines concern other aspects of FIGG, such as the prior use of CODIS and familial searching (where both lawful and feasible), as well as standards for how FIGG cases are to be managed by forensic laboratories (including quality control) and the vetting of vendor laboratories that process forensic evidence [10]. The NTVIC Guidelines also recommend that committees be established that are responsible for overseeing the ethical practice of FIGG and that would be composed of key stakeholders, including experts in criminal justice, racial injustice, civil and privacy rights, as well as families of victims and victims' rights advocates [10].

At present, only Maryland has passed a law requiring that practitioners of FIGG be licensed by an appropriate body, as set out in § 17-104(a)(2) of the *Criminal Procedure Code* [11]. The Office of Health Care Quality was to have established a licensing program for practitioners of FIGG before October 1, 2024, but there is no information at the time of writing that it has been able to do so [12]. The future may see the development of similar laws in other jurisdictions, and we can expect to see an increasing importance placed on professional certifications for FIGG practitioners, including adherence to professional standards and codes of ethics.

Each of the above bodies has issued professional standards and ethical guidelines relating to the practice of FIGG. This chapter will focus on those issued by the IGGAB, as they were developed specifically for practitioners of FIGG and are therefore intended for use in forensic applications such as human remains identifications and criminal investigations.

8.3 INDIVIDUAL AUTONOMY

Autonomy is a key value for assessing the ethics of FIGG. Promoting individual choice—for Reference Test Takers, relative matches, victims, and family members—is an important goal for FIGG practitioners, including law enforcement agencies who work with genetic genealogists. This includes autonomy over the choice to undergo genetic testing, as well as the ways in which genetic information is shared and used. Autonomy also encompasses bodily integrity, including control over the collection, storage, and use of biological samples.

8.3.1 Autonomy & Informed Consent

Informed consent is a long-standing principle in medical ethics that enshrines the liberal values of individual choice, autonomy, and dignity [13]. Informed consent has long played a key role in professional standards for genealogists. The BCG Code of Ethics requires genealogists to explain to persons providing data—including genetic data—what information they will collect, how the information is to be used and shared, as well as the benefits and consequences of that use [6]. Genealogists should advise test takers about their different options for sharing data and how this will affect their privacy, as discussed in 8.4, and explain that "there are never any guarantees of complete anonymity and privacy" [6].

Section 4 of the IGGAB Code of Ethics states that an AIGG guides their professional activities in a way that is consistent with individual autonomy [2]. Section 4.1(a) states that individuals who take a DNA test have the "right to make reasonable choices about how their genetic data will be used" [2]. The NTVIC Guidelines also require that FIGG practitioners respect the autonomy of those who use Genetic Genealogy Databases, as well as Reference Test Takers who are asked to assist in a FIGG investigation [10].

The IGGAB Code of Ethics requires AIGGs to provide certain information to individuals who are asked to provide genetic information to advance a forensic investigation, either through direct comparison or by uploading a genetic data file to a Genetic Genealogy Database (Reference Test Takers) [2]. AIGGs must advise Reference Test Takers about the possibility of unexpected discoveries, as set out in section 4.4(b) [2]. This may include, for example, unanticipated relatives, unknown medical conditions, unexpected biogeographical ancestry backgrounds, as well as misinformation about these topics.

AIGGs must obtain informed consent from Reference Test Takers, as set out in section 4.4(d) of the IGGAB Code of Ethics [2]. The NTVIC Guidelines also state that informed consent should be obtained except in those cases where it would jeopardize an ongoing investigation, for example, by causing a suspect to flee the jurisdiction [10]. "Informed consent" is defined in s. 3.8 of the IGGAB Professional Standards as "an individual or their legal proxy's written voluntary permission given with the understanding of the possible consequences and full knowledge of the possible risks and benefits of sharing their Genetic Data File" [3].

As part of informed consent, AIGGs have a duty to explain to Reference Test Takers how the investigating agency will access their data, the different privacy options available, and inform them that complete privacy and anonymity cannot be guaranteed, as per section 4.4(e) [2]. The BCG Code of Ethics requires that consent forms be in writing and that the genealogist obtain a signature, "freely given by the person providing the DNA sample or that person's guardian or legal representative" [6].

Ethical practitioners of FIGG therefore have a duty to obtain informed consent that is voluntary, free, and informed, and that must be in writing. Precedents for written consent forms for Reference Test Takers may be available from more experienced practitioners of FIGG. It would be desirable for the field to develop precedents embodying best practices for use in forensic applications, to invite public comment, and to make them available for practitioners.

8.3.1.1 Informed Consent: Free & Voluntary

To be fully informed, consent must be freely given, and this means freedom from coercion or undue pressure [13]. One example where consent to DNA testing was found to be coerced in a forensic context took place in Elgin County, Ontario, in the fall of 2013 [14]. A local woman was the victim of a violent home invasion sexual assault; she received medical attention and underwent a sexual assault kit which enabled a DNA profile of the suspect to be developed [14].

The victim also provided a description of her attacker that indicated that he was likely a migrant worker from one of the nearby farms [14].

The police decided to engage in a mass canvas of DNA samples from all migrant farmworkers in the area [14]. DNA canvases are a permissible law enforcement technique [15]. However, investigators must be sensitive to conditions that can render the technique unlawful or unethical.

In *Logan*, police officers asked the migrant workers to sign a consent form [14]. The consent form stated that the workers had the right to refuse to provide a DNA sample, and they had the right to consult with legal counsel [14]. The forms also stated that if there was no forensic match to their DNA, then the physical samples and associated information would be destroyed [14], which is a sound and ethical practice. Investigators collected DNA samples from 96 migrant workers, and police identified the individual responsible for the sexual assault [14].

However, the Ontario Human Rights Tribunal found that the migrant workers were vulnerable in ways that vitiated the consent they gave and rendered their compliance non-voluntary [14]. Under the terms of their temporary work visas, they were in danger of losing their jobs and being removed from Canada if they did not comply; employers (without the knowledge of the police) had advised the migrants that they would *not* be invited back to work next season if they did not comply with the investigation [14]. The Human Rights Tribunal also found that the DNA canvass was unnecessarily broad, in that police collected DNA samples even from migrant workers who clearly did not fit the description given by the victim [14].

Although this case concerned traditional forensic DNA and not FIGG, it imparts several valuable lessons for obtaining informed consent from Reference Test Takers. It is notable that the Human Rights Tribunal found that police had neither training nor a policy in place to guide them on how to obtain informed consent to DNA testing [14]. It would therefore be a wise practice for practitioners of FIGG—and for law enforcement who are using FIGG either in-house or through contractors—to develop and communicate sound policies regarding written informed consent for all DNA test takers.

These policies should account for any special vulnerabilities that Reference Test Takers may experience. Policies regarding informed consent should be communicated to all relevant stakeholders, to ensure that third parties do not engage in conduct or coercion that may vitiate free and informed consent.

8.3.1.2 Informed Consent: Risks & Benefits

Unexpected information concerning parentage, ancestry, and family history may arise in the course of any genealogical research, and not only genetic genealogy [16]. Russell notes that informed consent is the core of genetic genealogical ethics [16]. However, there is considerable debate about what informed consent means in the context of genetic genealogy, and what types of information should be included—particularly in the forensic context where the sharing of information may jeopardize an investigation.

The IGGAB Code of Ethics and Professional Standards [2, 3], like the BGC Standards [5, 6], imposes a duty of candor and honesty, as well as a duty to explain to Reference Test Takers the possibility of discovering unexpected genealogical information. As Russel states, genealogists must be candid with test takers and answer any questions they have about the test [16]. She states that genealogists "simply cannot secure informed consent if we misrepresent what we are asking the other person to do or what the risks are" [16].

The duty to advise Reference Test Takers includes a description of the terms of service and privacy policies of the Genetic Genealogy Databases used, as well as a description of any third-party tools being used [2, 16], what these tools are for, and how data is (and is not) shared via these tools [16]. It also includes a description of the different types of DNA tests being conducted, what they mean, and how they are shared [2, 16].

Informed consent should also include information about how the DNA sample will be stored, and what the individual needs to do to remove their information and destroy their sample. This information is normally found in the terms of service of the various Genetic Genealogy Databases.

Informed consent will also include information about how an individual can raise concerns and with whom. This includes information about how to lay a complaint regarding the investigative agency, as well as the relevant professional body that has issued a license to the forensic genealogist. Informed consent forms should include a copy of the relevant professional standards and codes of conduct in their appendices. Informed consent forms will typically have a statement advising the individual that they may seek out independent legal advice regarding the consent form and their rights with respect to the DNA testing.

DNA test takers will often initiate genetic genealogical research precisely because they want to know more about their ancestry and family history. Reference Test Takers in a FIGG investigation are not seeking out this information; they are being asked to provide their own genetic information to assist in an investigation that they have not initiated, and that is not for their personal benefit. Extra care should therefore be taken to obtain informed consent and to safeguard their information [13]. Further information on the privacy interests at stake is discussed in 8.4.

The IGGAB Code of Ethics requires that all Reference Test Takers be advised of the possibilities of discovering unexpected genealogical information, as well as the fact that complete anonymity and privacy cannot be guaranteed [2]. The discovery of unexpected paternity in one's family can be particularly upsetting [17]. Guerrini et al. have found that about 3% of users of Genetic Genealogy Databases discover unexpected paternity [18]. One review of studies reporting non-paternity found that, in North American and European societies, the non-paternity rate was about 1.9% [19].

A non-paternity event can disrupt a family over several generations. For example, Cuba Jiminez wrote in the Huffington Post about how a DNA test revealed that she had no Puerto Rican or Spanish ancestry [20]. Her grandfather turned out not to be the biological parent of her father [20]. She states that she had a few matches to second or third cousins, but "no real answers for the now increasingly long list of questions I had" [20]. It is unfortunate that many individuals who take DNA tests are not equipped to use the ancestry databases or third-party tools to resolve the questions they face, and training in the tools and techniques of genetic genealogy is only just beginning.

Sequoya Yiaueki wrote in *The Guardian* in 2018 about how profoundly his identity was disrupted when a DNA test showed that he had no Native American ancestry, as his father had long claimed [21]. The discovery of an unexpected ethnicity—which may well go along with the loss of a much-valued ethnicity—may be very troubling to some individuals.

Many of these problems can be resolved in forensic contexts by limiting the sharing of this type of information and by including statements in informed consent forms to the effect that this information will not be shared with the Reference Test Taker. This should be accompanied by an explanation, as set out in section 4.4(e) of the IGGAB Code of Ethics, that complete anonymity and confidentiality cannot be guaranteed [2], especially if a case is going to be adjudicated before a court.

Limiting the sharing of information with Reference Test Takers is one option, and a useful default when obtaining informed consent. There may be cases in which a relative match or a Reference Test Taker is adopted, and identifying their biological parentage is important for the resolution of the investigation. Some Reference Test Takers may want to know this information. Informed consent forms may be modified as needed to take account of situations such as this that arise in the course of a FIGG investigation.

There are many unresolved questions regarding the ethics of informed consent to reference testing in forensic cases. Should informed consent include information, for example, about the risk that the test taker or a relative could be charged with a crime? Should it include the risk that non-relatives could be charged with a crime? The risk that a perpetrator could victimize someone else if not apprehended? The *Logan* case discussed above is a reminder that coercion—even when seemingly mild or coming from third-party sources—may be enough to vitiate free and informed consent [14].

The best practice is therefore to keep information in informed consent forms aimed at the individual Reference Test Takers themselves, and the risks and benefits that may affect them personally [13]. Advising Reference Test Takers about benefits that may accrue to third parties (a family may find their missing loved one, a violent criminal may be apprehended before they harm someone else), or risks that could be faced by third parties (a relative could learn a family secret, or be charged with a crime) could place undue pressure on the individual, and will hamper their ability to give their consent freely and voluntarily.

8.3.1.3 Ethical Considerations for Reference Test Takers

Several ethical controversies have arisen over the ways in which Reference Test Takers have been approached and asked to undergo a DNA test, or to upload an existing genetic data file to a Genetic Genealogy Database to further a forensic investigation. Given the controversies these cases have caused, and the potential for public trust in FIGG to be disrupted, special care should be taken when approaching family members to conduct reference testing.

As discussed above, AIGGs have a duty to obtain written informed consent from Reference Test Takers [2, 3]. This includes, in section 4.4(b), the duty to inform the Reference Test Taker of the possibility they may obtain unexpected information [2]. The Code of Ethics also urges, in section 4.4(c), that AIGGs exercise discretion in their communications with test takers and DNA relatives [2].

For example, in their search to identify the remains of Baby Garnet, investigators approached one of the infant's relatives to provide some genetic information; the request was possibly that she upload an existing genetic data file from a Genetic Genealogy Database to GEDmatch to further the investigation [22, 23]. She posted a video to TikTok that caused quite a stir in the community when she claimed that police obtained her DNA profile from Ancestry.com, and that she was asked to provide the password to an account so that investigators could access it [22]. However, other reports cast doubt on these claims; court documents obtained by reporters showed that she did a DNA test on FamilyTreeDNA, which is accessible to law enforcement, and it may have been that she was being asked to upload this data file to GEDmatch, which is a common and permissible request of Reference Test Takers [22, 23].

It may be the case that Baby Garnet's relative did not understand what she was being asked to do, or the possibility that a family member—in this case her grandmother—would be identified as the mother of Baby Garnet and be charged in her death [22, 23]. Bioethicists debate the extent to which comprehension is necessary to find that informed consent has been properly given [13]. The obligation to provide informed consent will normally be discharged by providing information that is sufficient and easy to understand, and therefore provides a fair opportunity for the exercise of autonomy [13, 24]. Informed consent does not include an obligation to take extra steps to ensure the individual comprehends the information [13, 24].

Individuals have a corresponding obligation to read and to advise themselves of the information being provided to them before indicating their agreement [13, 25]. When an individual lacks the capacity to provide informed consent, then genealogists will obtain proxy consent from a legal representative [3, 26].

The NTVIC Guidelines recommend that FIGG practitioners not ask Reference Test Takers for their log-in credentials [10]. Instead, they may explain how to download a raw DNA file and upload it to an accessible Genetic Genealogy Database [10]. Alternatively, they may explain how to obtain a consumer DNA test and upload the results, or obtain a buccal swab from the individual [10]. The NTVIC Guidelines also recommend that only law enforcement communicate with Reference Test Takers and that they ask Reference Test Takers to sign a non-disclosure agreement to prevent them from disclosing information that would jeopardize an ongoing investigation [10]. These are sound best practices.

Investigators should be aware of the need to treat communications with Reference Test Takers with both sensitivity and caution. There is great value in obtaining written consent from individuals who are being asked to take a DNA test or to upload an existing profile to GEDmatch. It is excellent professional practice to document all such conversations in writing as they take place, in case investigators are later called upon to explain what transpired.

8.4 PRIVACY & CONFIDENTIALITY

Professional standards and codes of ethics require genetic genealogists to maintain the confidentiality of information and to respect the privacy of all individuals involved in genealogical research. This section will discuss some of the key ethical duties imposed on professionals with respect to maintaining the privacy of Reference Test Takers, DNA relatives, members of the public, and protecting the confidentiality of ongoing forensic investigations.

8.4.1 Adherence to Privacy Policies

Practitioners of FIGG have an ethical duty to know and to adhere to the privacy policies of the Genetic Genealogy Databases they are using.

The NTVIC Guidelines state that FIGG practitioners should respect both the terms of service of the Databases as well as the self-declared choices of their users [10]. The IGGAB Code of Ethics requires, at section 6.3(a), that AIGGs only upload and work with DNA profiles in accessible Genetic Genealogy Databases [2], which are defined in the IGGAB Professional Standards as those Genetic Genealogy Databases that give their users the explicit option to opt-in to FIGG or to keep their profiles private [3]. At the time of writing, FamilyTreeDNA [27, 28], GEDmatch [29–31], DNA Justice Foundation [32, 33], and Othram's DNASolves [34] are the Genetic Genealogy Databases that provide users the option to opt-in to law enforcement matching.

The BCG Genealogy Standard 57 requires respect for privacy rights [5]. This means that genealogists must inform individuals about the different privacy options available to them and may only share data from living persons with their written consent [5]. This applies to all types of data, and not only genetic data.

The BCG Code of Ethics also requires genealogists to explain the available privacy options, including how others can and cannot access their data [6]. This includes an explanation that complete privacy and anonymity can never be guaranteed [6]. As discussed above, genealogists must obtain written informed consent and then abide by those terms [6].

The IGGAB Code of Ethics states, in section 4, that an AIGG will conduct their professional activities in a way that is consistent with data privacy and respect for individual autonomy [2]. Section 4.1(b) places a duty on AIGGs to endeavor to protect the privacy of those whose genetic information is being evaluated [2]. Section 4.2(a) places a duty on AIGGs to endeavor to safeguard genetic data to which they have access, and to only use that data in ways that are consistent with the privacy choices those individuals have made in their account settings [2]. This includes genetic data files provided to an AIGG by an investigating agency, as well as genetic data that they would have access to in the form of relative matches.

In May of 2019, GEDmatch automatically opted all users out of law enforcement matching and required individual users to manually opt back in [35]. Between May 2019 and January 2021, some practitioners of FIGG used a loophole that enabled them to view the relative matches of a matching kit, even though the owners of some of those kits had "opted out" of law enforcement matching [36].

In August of 2023, Margaret Press, one of the founders of the DNA Doe Project, apologized for their use of the loophole [36]. She stated that, at the time, they did not consider how their failure to abide by the GEDmatch terms of service could weaken public trust in the use of genetic genealogy for forensic applications [36]. She stated that she should have reported the issue to GEDmatch right away so that the bug could be fixed [36].

The reports that some practitioners of FIGG used kits that had not opted in to law enforcement matching—and thus violated the database terms of use and used genetic information in a way contrary to the expressed wishes of its owners—were criticized by several genetic genealogists. Judy Russell, for example, decried the lack of consequences for those who had used the loophole and opined that it could harm ethical practitioners [37]. In response, GEDmatch fixed the loophole and undertook a system-wide assessment to detect and mitigate similar vulnerabilities going forward [38].

There have been reports of other violations of the terms of service of Genetic Genealogy Databases. In one incident, the FBI was accused of accessing the MyHeritage DNA database [39] to identify the body of a woman whose remains were found in the Badlands in Riverside County, California, in January of 1996 [40]. She had been sexually assaulted and murdered, likely only a few hours before being found [40].

An STR DNA profile was developed from her remains and entered into the California Missing and Unidentified Persons Index [40]. Her information was also entered into NamUs [40]. With no leads in the case, the Riverside County Regional Cold Case Homicide Team turned to Othram to develop a SNP profile so that a FIGG investigation could be commenced [40]. She was identified as 41-year-old Juana Rosas-Zagal [40].

However, in June of 2023, Larkin reported that an unidentified woman claimed that she had been contacted by detectives investigating the case, and they informed her that she may be related to the victim [39]. The woman provided a screenshot of her match list from MyHeritage, which indicated that she shared about 99.8 centimorgans of DNA over 11 shared segments with a woman identified as "Jane Dohue" [39]. The nickname appeared to be a clumsy attempt to disguise the identity of "Jane Doe" [39]. The DNA was not administered by Jane, but by an individual named Francis Rivarea [39].

Larkin reported that uploading an unidentified decedent's DNA to MyHeritage was both a violation of the terms of service of MyHeritage [39, 41], as well as the Department of Justice Interim Policy [41]. MyHeritage has long prohibited law enforcement use of its Genetic Genealogy Database [41]. The current privacy policy of MyHeritage, in force as of September 11, 2024, states that "MyHeritage prohibits law enforcement use of its DNA Services" [41].

The DOJ Interim Policy, effective on November 1, 2019, is a non-binding policy that sets out several guidelines for the use of FIGG [42]. The policy states that it is not legally binding and that it is not intended to limit the discretion of investigators and prosecutors [42]. The DOJ Interim Policy states, in section II, that it applies to criminal investigations over which the DOJ has exclusive or concurrent jurisdiction, and the investigating agency has lawful custody, control, or authority over the forensic sample [42]. It also applies to any criminal investigation that has received funding from the DOJ, any organization that has received funding for FIGG, and investigations in which a DOJ employee or contractor conducts FIGG [42].

The DOJ Interim Policy states, in section VII, that agencies using FIGG, "shall identify themselves as law enforcement to GG [genetic genealogy] services and enter and search FGG profiles only in those GG services that provide explicit notice to their service users and the public that law enforcement may use their service sites" [42]. The policy also states that SNP profiles entered into DNA databases by law enforcement for FIGG research should be configured to prevent their from being viewed by other users [42].

Therefore, if an investigator uploads a SNP profile to MyHeritage for FIGG searching, then this would violate both the terms of service of MyHeritage and the DOJ Interim Policy [41, 42]. If investigators do not identify themselves as law enforcement when they upload the profile, then this would also violate the DOJ Interim Policy [42]. Finally, if investigators make the profile available for viewing to relative matches, such as the women whose post was reported by Larkin, then this would also violate the DOJ Interim Policy [39, 42].

Although there are no consequences for a violation of the DOJ Interim Policy, which is not legally binding, incidents such as this weaken public trust and confidence in FIGG; they give the impression that law enforcement is not following ethical practices that are accepted in the field, and they are not accountable for ethical violations. For these reasons, the IGGAB has placed a duty on AIGGs to only use accessible Genetic Genealogy Databases, and to follow the terms of service and privacy policies of those databases [2, 3].

8.4.2 Eligible Cases & Qualifying Crimes

The IGGAB Code of Ethics requires, in section 6.2, that an AIGG will only use FIGG in cases where law enforcement "confirms that the goal is to identify a Subject by providing a Lead" [2]. A subject is, in turn, defined in the IGGAB Professional Standards as an individual "whose biological material is found in a probative location connected to the scene of a Qualifying Crime," in s. 3.15(a), or to identify an unknown decedent, as per s. 3.15(b) [3]. A "qualifying crime" is defined in section 3.13 as being that set out in the DOJ Interim Policy or an Accessible Genetic Genealogy Database [42].

The NTVIC Guidelines also state that FIGG should only be used in qualifying cases, and they also reference the DOJ Interim Policy's definition of qualifying crimes [10]. The guidelines recommend that FIGG be used in cases involving violent crimes, unidentified human remains, post-conviction relief, and where there is a significant risk to public safety [10]. The NTVIC Guidelines also recommend that familial DNA searching be used where this is permitted by law, and that STR profiles be developed and entered into CODIS databases whenever possible, prior to the use of FIGG [10]. Unlike the DOJ Interim Policy, the NTVIC Guidelines explicitly permit direct SNP-to-SNP comparison where it is not possible to obtain an STR profile [10]. This is an important update to the DOJ Interim Policy, and one that reflects advances in the available technologies since that policy was developed in 2019 [42].

Two statutes have been enacted to date that set out a definition for a qualifying crime. The Utah *Public Safety Code* at § 53-10-403.7 (1)(f) defines a "qualifying case" as an investigation into a violent felony or the identification of a missing or unknown individual, with no restriction that the individual must be deceased [43]. The Maryland *Criminal Procedure Code* 17-102(b)(1) states that FIGG may only be authorized when the crime "is the commission of, or the attempt to commit, murder, rape, a felony sexual offense, or a criminal act involving circumstances presenting a substantial and ongoing threat to public safety or national security" [11].

Montana has also passed a law concerning FIGG, but it includes no definition of a qualifying crime [44]. Instead, it states at § 44-6-104(a) and (b) that the government must not search consumer DNA databases unless a court has issued a warrant on a finding of probable cause, or the consumer has waived their right to privacy in the information [44].

8.4.2.1 Department of Justice Interim Policy

The DOJ Interim Policy sets out the definition of qualifying crimes in section 5, entitled "Case Criteria" [42]. FIGG may be used to investigate unsolved violent crimes—described in footnote 15 as:

> [...] any homicide or sex crime, including a homicide investigation during which FGGS is used in an attempt to identify the remains of a suspected homicide victim. It also includes other serious crimes and criminal offences designated by a GG [genetic genealogy] service for which investigative use of its service by law enforcement has been authorized for that service. [42]

The Interim Policy also states that, "in addition," a prosecutor may authorize the use of FIGG:

> [...] for violent crimes or attempts to commit violent crimes other than homicide or sexual offences (while observing and complying with all requirements of this interim policy) when the circumstances surrounding the criminal act(s) present a substantial and ongoing threat to public safety or national security. [42]

Therefore, a qualifying crime is expansively defined in the DOJ Interim Policy—authorizing FIGG for homicide, any sexual offenses, and other violent crimes (including attempts) that present a "substantial and ongoing threat" to the public—and gives the prosecuting attorney a degree of discretion to authorize the use of FIGG for any violent crime that poses a significant risk to the safety of the public.

Notable, too, is what the policy leaves out. It does not mention the use of FIGG for the identification of unknown decedents who are not victims of homicide—and which form a substantial part of all human remains identifications cleared by FIGG [45]. The IGGAB Professional Standards do authorize, in s. 3.15(b), the use of FIGG for identifying human remains irrespective of whether they are thought to be a homicide victim [3].

8.4.2.2 The IGGAB Professional Standards

The IGGAB Professional Standards define a "qualifying crime" in section 3.13 as being that set out by the DOJ Interim Policy, and which is incorporated by reference into the standards, and by the terms of the Genetic Genealogy Databases, which are accessible for law enforcement use [3].

The IGGAB Code of Ethics requires that AIGGs comply with the terms of service (including any privacy policies) of Genetic Genealogy Databases, as set out in section 6.3 [2]. This means that AIGGs may only use Genetic Genealogy Databases whose terms of service permit use by law enforcement, as per section 6.3(a) [2]. At the time of writing, this includes Family Tree DNA, GEDmatch, and DNA Justice, and DNA-Solves [27–33]. AIGGs must also know the terms of service of the genetic genealogy databases they are using and adhere to those terms—as set out in section 6.3(b) [2].

Section 6.3(b) of the IGGAB Code of Ethics requires that AIGGs abide by the self-declared privacy settings of the users of those databases [2], and Section 6.3(d) places a duty on AIGGs to report any issues with Genetic Genealogy Databases that could impact the privacy of users [2]. These two provisions seem specifically tailored to prohibit the use of loopholes, as took place in 2019 with GEDmatch and discussed above. They require AIGGs to report any bugs they become aware of that can affect the privacy of database users.

8.4.2.3 Genetic Genealogy Database Terms of Service

Genetic Genealogy Databases that are accessible for FIGG set out the type of cases that are eligible for FIGG in their terms of use and privacy policies. Like all terms of service, these change over time, can be expected to change in the future, and when questions are raised over whether an offense is eligible for FIGG, this has created ethical controversies. It is worth noting that, since policies and terms of service change over time, adherence to those policies must be assessed according to the policies then in place.

8.4.2.3.1 DNA Justice The DNA Justice Foundation is a non-profit organization that has designed a database that law enforcement agencies may use for FIGG investigations in eligible cases [32]. Members of the public may download their raw genetic data files from other Genetic Genealogy Databases and upload them to DNA Justice, including a GEDcom if the user wishes [32]. Unlike FamilyTreeDNA and GEDmatch, the DNA Justice database is private and is only accessible to approved investigating agencies [32]. Like FamilyTreeDNA and GEDmatch, investigating agencies will only have access to the relative matches, and not the user's genetic data files themselves [32].

DNA Justice defines an "eligible case" as one that seeks to identify human remains, or to investigate the violent crimes of "murder, nonnegligent manslaughter, rape/sexual assault, abduction, robbery, aggravated assault, terrorism, imminent threats to public safety—and any attempts at the first three crimes" [32]. Like GEDmatch [30], discussed below, it prohibits its use to identify fetal remains and stillborn children when used to implicate the birthing parent in a criminal offense [32].

DNA Justice also clarifies that information about forensic investigations will not be shared with database users. This includes, for example, information a user may not know about their family, or the discovery that a user's biological father is different from what was expected [32].

8.4.2.3.2 FamilyTreeDNA The current FamilyTreeDNA Privacy Statement, effective as of August 6, 2024, states in section 5.A that FamilyTreeDNA may collaborate with law enforcement agencies to conduct FIGG searches [27]. A user's information will be viewable through a law enforcement account if the user has opted in to relative matching, has opted in to FIGG matching, and the user is a relative match to a profile uploaded by a law enforcement account [27].

All users may enable and disable the relative matching feature, including FIGG matching, in their privacy settings [27]. Where enabled, relative matches will see "the markers they share with you, the degree of relatedness between you, and some of your Personal Information, such as your name, email, and other profile information" [27]. FamilyTreeDNA automatically opts all users into relative matching, requiring users to go into their account settings and opt-out [46].

The current FamilyTreeDNA Law Enforcement Guide is effective as of April 25, 2024 [28]. FamilyTreeDNA may grant access to a law enforcement agency to open a law enforcement account [28]. FamilyTreeDNA may grant permission to law enforcement to use their account for qualifying cases, which include the identification of human remains, whether or not the individual is suspected to be the victim of a homicide, or to identify individuals suspected of committing a violent crime [28].

"Violent crime" is defined as "any homicide (including murder or non-negligent manslaughter), any sex crime (including rape), or any of the following serious crimes or criminal offenses: (1) armed robbery, (2) aggravated assault, or (3) any other violent crimes as defined by the FBI Uniform Crime Reporting (UCR) Program" [28]. The Law Enforcement Guide includes attempts to commit a violent crime, as well as "when the circumstances surrounding the criminal act or acts present an ongoing threat to public safety or national security" [28].

In turn, the FBI's Uniform Crime Reports include within their data collection for violent crimes homicide, rape, robbery, and aggravated assault [47]. The current definition of "rape" in the UCR, adopted in 2013, is "[p]enetration, no matter how slight, of the vagina or anus with any body part or object, or oral penetration by a sex organ of another person, without the consent of the victim" [48].

The FBI UCR definition of violent crime is therefore more restrictive than the current FamilyTreeDNA definition. The FBI UCR definition is also more restrictive—particularly for sex offenses—than that of the National Incident Based Reporting System (NIBRS) that has been developed to replace the UCR's Summary Reporting System [49].

8.4.2.3.3 GEDmatch Pro GEDmatch's Privacy Policy, effective as of November 22, 2024, permits law enforcement use to identify the perpetrator of a violent crime, defined as, "murder, nonnegligent manslaughter, aggravated rape, robbery or aggravated assault" or to identify human remains [29].

Practitioners of FIGG upload the genetic data files they are investigating to GEDmatch PRO. The Terms of Use of GEDmatch PRO, effective as of October 28, 2024, permit the use of their services to identify the perpetrator of a violent crime, where "violent crime" is defined as "murder, nonnegligent manslaughter, aggravated rape, robbery, or aggravated assault" [30]. The database may also be used to identify human remains (and not only those thought to have died by homicide), but may not be used to identify fetal remains or the remains of stillborn children [30].

The term "aggravated rape" does not align with the definition of sexual offenses used in the above definitions and does not have a consistent definition across jurisdictions. Some jurisdictions, such as Canada and the District of Columbia, for example, no longer use the term "rape" to define any sexual offense in their penal codes [50, 51]. As described above, the FBI UCR defines rape as the "[p]enetration, no matter how slight, of the vagina or anus with any body part or object, or oral penetration by a sex organ of another person, without the consent of the victim" [48], and does not include a definition of aggravated rape.

The current use of the term "aggravated rape" is therefore difficult to map onto the various definitions of sexual offenses as set out in modern penal codes, several of which do not use the term "rape." GEDmatch's current definition here is out of step with the way in which sex offenses are normally defined and may be unnecessarily confusing for practitioners of FIGG to work with.

GEDmatch's current Terms of Service, effective November 22, 2024, allow users to customize their privacy settings [31]. Users may choose from four levels of privacy settings, and users may change their privacy settings at any time [31]. The "private" setting means that kits will not be available for many relative matching or comparisons with any other kit, and not only kits uploaded by law enforcement in GEDmatch PRO [31].

GEDmatch's "opt-in" setting will compare the user's kit with each of the other kits in GEDmatch's database [31]; this includes comparison with law enforcement kits that have been uploaded to GEDmatch PRO to identify perpetrators of violent crimes and unidentified human remains [31].

The third privacy choice is that of "opt-out," which will allow the user's kit to be compared with all other kits in the database, except kits uploaded by law enforcement to identify perpetrators of violent crimes; this setting permits the user's kit to be compared with law enforcement kits to identify human remains [31].

The fourth privacy option is "personal research," which allows the user's kit to be compared with each of the other kits in the database but will not permit the user's kit to be shown in the match results reported for other users—including kits uploaded by law enforcement in both human remains and violent crime investigations [31]. This choice effectively allows users to see their own relative matches, but does not allow other users (including law enforcement) to see them as a match, or to contact them [31].

8.4.2.4 Controversies Over the Definitions of "Qualifying Crime"

The current privacy settings of GEDmatch have a long and storied history. This has given rise to several controversies over the various definitions of "qualifying crime" in use, and whether the use of FIGG to clear some cases has adhered to the terms of use of the Genetic Genealogy Databases being used.

8.4.2.4.1 The Centerville Aggravated Assault The first controversy arose in Centerville, Utah, where detectives used FIGG to identify a suspect in a November 2018 attack in a local church [52]. The victim, then 71 years old, was alone in the church playing the organ when an unknown assailant broke in and choked her until she lost consciousness [53]. Investigators collected the suspect's from underneath a broken window [53]. The suspect was 17 years old; he was prosecuted as a youth for aggravated burglary, aggravated assault, and mischief, with much of the court file being sealed to protect his privacy [53].

Detectives with the Centerville Police Department initially approached Parabon Nanolabs to conduct the FIGG investigation, but were told by Parabon that (at that time) the GEDmatch terms of service only permitted FIGG searching for the crimes of homicide and sexual assault [52, 54]. Detectives did not want to end the investigation there, and so they approached those in charge of GEDmatch [52, 55]. At the time, the site was still being managed by genetic genealogy hobbyists Curtis Rogers and John Olson, who founded the site in 2010 [55].

Rogers was prevailed upon to give police his permission to use the GEDmatch database in the Centerville attack, because they convinced him that the perpetrator was dangerous, and that public safety was at risk so long as he remained at large; the victim herself was in fear of her life [52, 55]. Parabon's genetic genealogists identified the perpetrator in late March 2019 [52].

Several critics decried the use of FIGG in this way [52]. Since this was the first time that FIGG had been used to clear a case of aggravated assault, critics claimed that this was evidence of a slippery slope whereby GEDmatch would be used for more and more minor offenses [52]. However, these fears have not materialized; as of July 31, 2024, there have been only 2 out of 358 perpetrators recorded as being identified by FIGG and whose most serious crime was an aggravated assault [45].

There have been no perpetrators identified through FIGG where the offense was a less serious charge of assault than aggravated assault [45]. Aggravated assault is itself a serious offense which carries the possibility of a lengthy jail term in most jurisdictions and—while definitions vary across jurisdictions—normally includes as part of its definition that the perpetrator has seriously injured the victim or placed their life at risk [e.g. 50].

Critics also contended that the investigation failed to respect the GEDmatch terms of service in place at the time [54] and was done without providing notice to users or affording them an opportunity to consent [56]. As a result of these criticisms, GEDmatch changed its terms of service in May of 2019 to automatically opt all users out of law enforcement matching [35, 56]. Users had to manually change their privacy settings to permit law enforcement to access their profiles for FIGG searches [35, 56]. These terms were modified in GEDmatch's next update to their terms of service and privacy policy on December 30, 2021, to automatically opt all users in to law enforcement matching for human remains investigations [57].

This sudden change in the GEDmatch terms of service was done without warning to investigators, and without consultation [58]. As a result, many ongoing FIGG investigations were stalled, and the data being used in ongoing academic research was abruptly lost [58]. This created much ill feeling in the field, satisfying neither practitioners of FIGG nor its critics [56, 58].

It was after this change in the GEDmatch terms of service that some FIGG practitioners began to use the loophole described above [36]. This incident offers an important lesson for the development of ethical norms within the field: rules that are handed down without consultation or broad acceptance may be perceived as one-sided and illegitimate—and are therefore less likely to be respected and complied with.

Commentators on both sides of the debate feared that the new automatic opt-out rule would lead to an increase in police obtaining warrants to access Genetic Genealogy Databases, including the massive database of Ancestry.com [56, 59]. This fear has not materialized. To date, there is a public record of only two warrants authorizing access to a Genetic Genealogy Database, and both were issued in 2019 after GEDmatch's changes to its terms of service [59, 60]. One warrant was issued by the Ninth Judicial Circuit Court in Florida in July of 2019, and it permitted police in Orlando to search the entire GEDmatch database [59].

Experts, such as Erin Murphy of NYU, stated that this was a "huge game changer," and would encourage police to obtain similar warrants to access other, larger, databases such as 23andMe and Ancestry.com [59]. Murphy told the *New York Times*, "I have no question in my mind that if the public isn't outraged by this, they will go to the mother lode: the 15-million-person Ancestry database" [59].

News reports indicate that the Florida warrant was issued because it was one of the cases impacted by GEDmatch's change to the automatic opt-out policy—which is to say that the suspect's kit had already been uploaded to GEDmatch, but relative matches which were recorded prior to the opt-out policy were lost when that policy was implemented [61]. This evidence established probable cause that there were in fact relative matches in the database [61].

Ancestry reported a warrant being issued to search their DNA database in their 2019 Transparency Report [60]. Ancestry states that it challenged the warrant on jurisdictional grounds and that no user data was provided in response to the warrant [60]. News reports indicated that the warrant was issued by a court in Pennsylvania, but did not state whether the warrant was issued in connection with a FIGG investigation [62]. There have been no further reports of warrants to search Genetic Genealogy Databases since these two instances were reported in 2019.

In January of 2021, the remains of 14-year-old Wendy Stephens—the youngest known victim of serial killer Gary Ridgway—were identified by the DNA Doe Project using FIGG [63]. She had likely been deceased for a year when her remains were found in March of 1984 in a wooded area in SeaTac [63]. Cairenn Binder, who led the team of volunteers at DNA Doe, reported that one of Stephens' parents had uploaded their DNA to GEDmatch in hopes of finding Wendy, or any children she may have had [63]. However, they were not aware of the 2019 change in GEDmatch's terms, and did not know that they had to opt back in to law enforcement matching [63]. Their profile was therefore not accessible to DNA Doe, who took much longer to make the identification using more distant relative matches [63].

In December of 2021, GEDmatch again changed its terms to give users the option to opt out of law enforcement matching, but all profiles would be used for the identification of unidentified human remains; users who wished to opt out of these types of FIGG investigations would have to change their privacy setting to "private," where it could not be used for any relative matching [57]. However, this afforded no option for users who wished to opt out of all law enforcement matching, even for human remains identifications. This is now permitted by the privacy options available in the current GEDmatch Terms of Service that give users the opportunity to remove their kits from all forensic investigations, and to keep them private from other users [31].

8.4.2.4.2 Stalking & Uttering Threats Another case that presented an ethical controversy surrounding the definition of a "qualifying crime" is the FBI's use of FIGG to identify James David Rogers as the individual who, over the course of 12 years, mailed dozens of threatening letters to actress Eva LaRue [64]. The writer threatened to rape and kill LaRue and her young daughter [64].

The FBI obtained a DNA profile from the envelopes [64]. In September of 2022, Rogers pled guilty to federal charges, including two counts each of mailing threatening communications and stalking, and one count of making threats by interstate communication, and he was sentenced to 40 months in prison [64]. At his sentencing hearing, he took responsibility for his actions, and he apologized to his victims [64].

Offenses related to stalking and uttering threats are offenses that may pose a substantial and ongoing risk to public safety. If so, then they are within the definition of a qualifying crime for DNA Justice [32], FamilyTreeDNA [27, 28], and the DOJ Interim Policy [42]—and therefore the IGGAB Professional Standards [3].

These examples illustrate how problems may arise because of inconsistent definitions across the different Genetic Genealogy Databases and policies in place. In particular, the more restrictive definition used by GEDmatch versus the DOJ Interim Policy may continue to create conflict [30, 42]. Investigators—such as the FBI—who are covered by the DOJ

Interim Policy may be authorized to use GEDmatch in ways that accredited AIGGs are not. A uniform definition would prevent these situations from arising.

8.4.2.4.3 The College Station Sexual Assaults Christopher Quinn Williams, then 39 years old, was sentenced in August of 2022 to life in prison for burglary and attempted sexual assault for an offense that took place in College Station, Texas, in April of 2018 [65]. This offense was part of a series of at least eight similar offenses [65, 66]. He broke into the victim's apartment while she was asleep; she awoke with one of his hands on her mouth, and he was trying to put his other hand in her underwear and was grabbing her genitals [66]. A DNA profile of the suspect was obtained from semen collected from the victim's clothing [66]. Some of the victims were later harassed through text messages referencing the attacks [66].

Williams was convicted under § 30.02(a)(1) (burglary of a dwelling house with intent to commit a felony), namely § 22.011 (sexual assault) of the *Texas Penal Code* [67], and these convictions were upheld by the Texas Court of Appeals in 2023 [66]. Williams received a life sentence for these offenses [66]. Under articles 62.001(5)(A) and (D) of the *Texas Code of Criminal Procedure*, these offenses require registration as a sex offender [68]. The offenses for which Williams was convicted are therefore sexual offenses according to Texas law.

However, several reports about the case either eliminated or downplayed the sexual nature of Williams' offenses. For example, the *LA Times* referred to this as a "burglary" case, stating that "police met search guidelines by classifying a case as sexual assault but after an arrest only filed charges of burglary" [61]. The article omits to note that Williams was charged with burglary with intent to commit a sexual assault [61]. The author also minimized the offenses, stating that Williams intruded into the women's bedrooms, "stealing nothing but sometimes touching a sleeping woman while masturbating" [61].

The *LA Times* claimed that the use of FIGG in this case violated GEDmatch's terms of service [61]. This offense took place in April of 2018, and Williams was arrested on December 8, 2018 [69]. The terms of service that were in effect at the time permitted GEDmatch to be used by law enforcement for the violent crimes of "homicide and sexual assault" [54].

In May of 2019, GEDmatch changed the definition of a qualifying crime from "sexual assault" to the current definition of "aggravated rape" [35]. There is no information as to whether GEDmatch restricted the use of their database only to cases of aggravated rape because of the criticism they received over its use in the Williams case [61].

Williams' offense would certainly fit each of the other organizations' definitions of a qualifying crime. DNA Justice permits FIGG for sexual assaults, attempted sexual assaults, and imminent threats to public safety [32]. FamilyTreeDNA permits the use of FIGG for "any sex crime" as well as "ongoing threats to public safety" [28]. The DOJ Interim Guidelines [42]—and therefore the IGGAB Professional Standards [3]—would also permit the use of FIGG for any sex crime, as well as for substantial and ongoing threats to public safety.

8.4.2.5 Missing Persons & Living Does

None of the above definitions explicitly permits the use of FIGG to identify missing persons and living does. Living cases are unusual, and only a few have been cleared to date [45]. One instructive example is that of "Riverside Doe." Riverside Doe is a non-verbal cognitively disabled adult who was found unconscious and injured near a homeless encampment in Riverside, California, on Christmas Eve in 2018 [70]. Because of his cognitive disabilities, he was not able to identify himself; it seems clear that he was not being looked after, and no one came forward to claim him, even after a campaign to reach out to the public [70].

Police used FIGG to identify Riverside Doe [70]. Would this case fit the DOJ Interim Policy definition of an eligible crime? In this case, the individual was cognitively disabled and lacked the capacity to consent to DNA testing; proxy consent was provided by the County of Riverside Public Guardian [70].

This case may fit the definition of an eligible case that a prosecutor may approve under the DOJ Interim Policy, since it involves criminal offenses relating to neglect, and which had already placed the safety of Riverside John Doe at risk. Since the DOJ Interim Policy's definition of a qualifying crime [42] has been adopted by reference into the IGGAB Professional Standards, at section 3.13 [3], then an AIGG would also be permitted to work on such a case.

The Maryland *Criminal Procedure Code* specifically addresses missing persons, at § 17-101.1(2), where it states that its restrictions do not apply to the use of FIGG to identify a missing person—but only when a missing person report has been filed with a law enforcement agency and the person's whereabouts are unknown [11]. It appears that the Maryland law would not authorize the use of FIGG to identify living does like Riverside Doe, whom no one reported missing [11].

This is perhaps an undesirable result, since there are no practice standards or ethical norms that would prohibit genealogists in non-forensic contexts from working on such a case on behalf of a private individual or a public guardian. The absence of rules permitting the use of FIGG to identify living does may be an oversight, perhaps due to the rarity of these cases. There seems to be no valid ethical reason to prevent the use of FIGG to identify living persons and other missing persons.

The Utah *Public Safety Code* at § 53-10-403.7 (1)(f)(ii) defines a "qualifying case" as an investigation of the identity of a missing or unknown individual [43]. It would therefore appear to authorize investigations into the identity of a living doe like Riverside Doe. This definition should be considered by those seeking to formulate criteria for qualifying cases that include missing persons and living does.

8.4.2.6 The Future of the Definition of "Qualifying Crime"

Where the terms of use and privacy policies of a Genetic Genealogy Database set out a definition of an eligible case that is more restrictive than that set out in the DOJ Interim Policy, then an AIGG must adhere to the definition as set out in the database terms. To do otherwise would be to risk being in violation of section 6.3 of the IGGAB Code of Ethics, which requires compliance with Genetic Genealogy Databases' terms of service [2].

However, the differences between the definitions of eligible cases and qualifying crimes as set out in the DOJ Interim Policy [42] and the various accessible Genetic Genealogy Databases [27–33] have the potential to create confusion for practitioners of FIGG as to what offenses are permitted—particularly for sexual offenses. The various definitions of eligible cases and qualifying crimes are set out in Table 8.1.

First, the differences in the way that crimes are defined create confusion and controversy over the correct interpretation of qualifying crimes. Second, it has the potential to place AIGGs in a difficult position—one in which prosecutors are authorizing them to conduct an investigation for an offense that fits the definition of a qualifying crime in the DOJ Interim Policy [42] but that does not clearly fit the definition of a qualifying crime under the terms of use of a given Genetic Genealogy Database [27–33]. This confusion is most likely to arise for sex offenses, as well as offenses that pose a threat to public safety.

8.5 HUMAN DIGNITY

The importance of human dignity in FIGG practice encompasses many of the ethical issues discussed above, as well as some that are not covered elsewhere. Here, I would like to touch upon some common issues that may arise in the course of FIGG. A more complete discussion of these important topics is, unfortunately, outside the scope of this chapter, but I hope to at least familiarize the reader with these issues and help to keep them in mind in the daily practice of FIGG.

TABLE 8.1 Definitions of Eligible Cases and Qualifying Crimes

Case Type	DOJ Interim Policy [42]	IGGAB Professional Standards [3]	DNA Justice Foundation [32]	Family Tree DNA [28]	GEDmatch PRO [30]	Maryland Code of Criminal Procedure, Title 17 [11]	Utah Public Safety Code §403.7 [43]
Unidentified decedents who are victims of homicide	Identification of the remains of a homicide victim	An individual who is deceased and whose identity is unknown	Unidentified human remains	Unidentified decedents, regardless of whether they are a suspected homicide victim	Unidentified human remains		Missing or unknown individual (403.7(1)(f)(ii))
All unidentified decedents		An individual who is deceased and whose identity is unknown	Unidentified human remains	Unidentified decedents, regardless of whether they are a suspected homicide victim	Unidentified human remains	Identification of all deceased individuals (17-101.1)	Missing or unknown individual (403.7(1)(f)(ii))
Fetal tissues and stillborn remains			Specifically prohibited where it is done to implicate the birthing parent in a crime		Specifically prohibited		

(*Continued*)

TABLE 8.1 (Continued)

Case Type	DOJ Interim Policy [42]	IGGAB Professional Standards [3]	DNA Justice Foundation [32]	Family Tree DNA [28]	GEDmatch PRO [30]	Maryland Code of Criminal Procedure, Title 17 [11]	Utah Public Safety Code §403.7 [43]
Missing persons/ Living does						Identification of persons who are the subject of a missing person report and whose whereabouts are unknown (17-101.1)	Missing or unknown individuals (403.7(1)(f)(ii))
Murder	Any homicide	Any homicide	Murder and attempts, but excluding negligent manslaughter,	Murder and attempts, but excluding negligent manslaughter	Murder, but excluding negligent manslaughter	Murder and attempts (17-102)	A violent felony (403.7(1)(f)(i))
Rape	Any sex crime	Any sex crime	Rape, and attempts	Rape, and attempts	Aggravated rape	Rape and attempts (17-102(b)(1))	A violent felony (403.7(1)(f)(i))
Sexual offenses	Any sex crime	Any sex crime	Sexual assault and attempts	Any sex crime, and attempts		Felony sexual offenses and attempts (17-102(b)(1))	A violent felony (403.7(1)(f)(i))

Threat to public safety	Violent crimes where the circumstances present a substantial and ongoing threat	Violent crimes where the circumstances present a substantial and ongoing threat	Imminent threats to public safety	Circumstances present an ongoing threat		Substantial and ongoing threat to public safety (17-102(b)(1))
Threat to national security	Violent crimes where the circumstances present a substantial and ongoing threat	Violent crimes where the circumstances present a substantial and ongoing threat		Circumstances present an ongoing threat		Substantial and ongoing threat to national security (17-102(b)(1))
Other	Other serious crimes designated by a Genetic Genealogy Database as being authorized for law enforcement use	Other serious crimes designated by a Genetic Genealogy Database as being authorized for law enforcement use	Abduction, robbery, aggravated assault, terrorism, but not attempts	Armed robbery, aggravated assault, and other violent crimes as defined in the FBI UCR, including attempts	Robbery, aggravated assault	

Note: This table sets out some of the key laws, policies, and terms of service/privacy policies of accessible Genetic Genealogy Databases, that define what cases are eligible for FIGG.

8.5.1 Dignified Treatment: Honesty & Candor

The professional obligations of honesty and candor are relevant to the ethical goal of respecting human dignity, as are the duties of civility and courtesy. This includes the dignified treatment of family members of decedents, as well as family members of those accused of criminal offenses or other illegal or immoral conduct. It also includes the dignified treatment of human remains and victims of crime, and a particular respect for their dignity—promoting which are key goals of the field as a whole.

The above section discussed some of the ethics of communicating with Reference Test Takers and DNA relatives, including the importance of respecting their autonomy, obtaining informed consent, and abiding by users' self-declared privacy settings. Here, I would like to address the issue of whether investigators may provide false information to Reference Test Takers and DNA Relatives. Such practices will often be lawful, and the evidence obtained through ruses will often be admissible as evidence in criminal proceedings [see e.g. 71,72]. However, because these practices have the potential to weaken public trust in the use of FIGG, investigators should use other means of protecting the confidentiality of ongoing investigations.

One such controversy arose in the course of the investigation into the October 2001 murder of Christine Franke [71]. A DNA profile of the suspect had been developed from semen collected from Franke's body [71]. Investigators then used FIGG to identify Benjamin Holmes, who was convicted in February of 2023 by a jury of first-degree murder in the death of Franke [71].

In October of 2018, detectives from Orlando, Florida, approached Eleanor Holmes for further information [72]. Eleanor Holmes told reporters that police were looking for her assistance to identify an unknown decedent who may have been a relative of hers [72]. She knew that police had already obtained her sister's DNA; she gave them a swab, thinking that it might help identify a missing niece [72]. Her son, Benjamin, was arrested a few days later for Franke's murder [72].

Eleanor felt that the police had deceived her, and she told reporters that she would not have provided her DNA had she known that the police were investigating a relative for a criminal offense [72]. Other—more distant relatives—were glad to have assisted in the investigation [73]. It is not surprising that close versus distant relatives would have different interests at stake in a FIGG investigation.

Undercover ruses to obtain DNA in the course of criminal investigations have been found lawful by courts in both the US and Canada [e.g. 74, 75]. It is therefore unlikely that evidence obtained by the use of ruses and undercover police operations would be excluded in subsequent criminal proceedings. However, as in *Logan* [14], it may give rise to civil complaints regarding violations of privacy, human rights, and civil liberties. Unlike police, AIGGs have a duty to obtain informed consent, as discussed above. This is a best practice, and one that should be respected by all investigators who are asking for genetic information from Reference Test Takers and DNA relatives.

The DOJ Interim Policy permits DNA from Reference Test Takers to be obtained with informed consent, or surreptitiously after obtaining a warrant in those cases in which this would compromise an ongoing investigation [42], as does the Maryland *Criminal Procedure Code* [11]. This should be considered a best practice in the field as a whole.

Where investigators cannot provide information to Reference Test Takers or DNA relatives without compromising the integrity of an investigation, they must state this and permit the individual to decide for themselves if they would like to assist. Alternatively, they may seek a warrant authorizing the search or seizure. Investigators and practitioners of FIGG should not provide false information that may violate the individual's dignity or compromise their ability to exercise their autonomy.

8.6 PROFESSIONAL INTEGRITY

Much of the work of professional bodies such as the IGGAB, the BCG, and CAFG is devoted to ensuring that those who hold professional credentials are held to high standards of competence in conducting genealogical research. These standards include requirements for periodic recertification and/or ongoing professional development.

8.6.1 The IGG Proof Standard

Conducting high-quality genetic genealogical research includes meeting the genealogical proof standard. The BCG genealogical proof standard has five components, each of which is described at length in the BCG's Genealogical Standards [5]. The GPS requires (1) reasonably exhaustive research, to reduce the likelihood that there exists undiscovered evidence that contradicts the research conclusions; (2) Complete and accurate source citations; (3) A critical analysis of the best available evidence; (4) a resolution of any conflicting evidence; and (5) a soundly reasoned conclusion that clearly explains how the evidence leads to the conclusion [5].

The IGGAB Professional Standards [3], at section 6.1, incorporate by reference the best practices outlined in the BCG Standards [5], and with which all professional genealogists will be familiar. However, the IGG Proof Standard is not the same as the BCG genealogical proof standard, which must be adapted for use in the forensic context.

The IGGAB Professional Standards defines the IGG Standard of Proof in section 6.1, which is met when the AIGG generates "Leads that may suggest the identity of the Subject" and is satisfied when the AIGG produces "one or more Leads such that further investigation is justified" [3]. The IGG Standard of Proof "shall ensure that an AIGG does not waste an Investigating Agency's resources nor subject individuals to undue suspicion or scrutiny" [3]. Leads produced by an AIGG are not themselves intended to be conclusive proof of identity [3]. This is similar to the DOJ Interim Policy that states that a "suspect shall not be arrested based solely on a genetic association generated" through FIGG [42].

The IGGAB Professional Standards state that AIGGs will adhere to the best practices set out in the BCG GPS, but that they will adapt this to the specific context of forensic work [3]. For example, this includes maintaining adequate documentation of work, such as through a research log or repository where genealogical research and sources are documented, but does not necessarily require the same standard of citing each genealogical source relied upon in an investigation [3].

8.6.2 Confidentiality & Security of Investigative Information

The IGGAB Code of Ethics places a duty on AIGGs to protect the confidentiality and security of investigative information. Section 4.3(a) requires AIGGs to protect "confidential case information as mandated by the Investigating Agency or any other legal entity" [2]. The NTVIC also states that strict confidentiality is required for all data and identifying information [10].

Section 4.1(d) of the IGGAB Code of Ethics requires AIGGs to maintain strict confidentiality with respect to the identities of individuals involved in investigations, as well as their genetic data, and permits sharing information "only with authorized parties and within the boundaries of the law" [2]. This includes sharing case information with DNA relatives and Reference Test Takers only under the direction of the investigating agency, as per section 4.3(f), and communicating with the investigating agency in a manner that protects confidential case information, as per section 4.3(a) [2].

AIGGs must also maintain high standards of data security. This includes a duty to store and transmit genetic data using "robust data security measures to minimize the possibility of unauthorized access or use," as set out in section 4.2(c), and a duty to develop and follow "clear policies for the retention, backup, anonymization and disposal of Genetic Data Files in compliance with relevant laws and regulations," as per section 4.2(h) [2].

8.6.3 Professional Competency

The IGGAB Professional Standards also impose requirements for knowledge and professional competency. These are set out in section 6.2 and summarized in Table 8.2 [3]. These standards will form the basis of an exam that candidates must pass in order to become accredited as AIGGs, as set out in section 5 [3]. AIGGs will be required to maintain their professional competency through continuing education, as set out in section 6.3(c) [3].

TABLE 8.2 Description of the Professional Competencies Required for an AIGG as Set Out in Section 6.2 Competency Standards of the IGGAB Investigative Genetic Genealogy Professional Standards and Accreditation Requirements, effective April 2024 [3]

Section	Description of Competency
6.2.1	Fundamentals of investigative genetic genealogy, including common misconceptions about the field
6.2.2	Fundamentals of DNA evidence, including STR typing, CODIS, and SNP sequencing, and forensic evidence collection
6.2.3	Fundamentals of genetic genealogy, including principles of inheritance, recombination, biogeographical ancestry, and relative matching
6.2.4	DNA SNP processing and bioinformatics, including types of DNA processing, imputation, and how DNA contamination and degradation affect case outcomes
6.2.5	Ethical and privacy standards, including knowledge of the IGGAB Code of Ethics and Professional Standards
6.2.6	Basics of using GEDmatch and GEDmatch PRO, including terms of service, restricted uses, and associated tools
6.2.7	Basics of using other accessible databases, including terms of service, restricted uses, and associated tools
6.2.8	Ability to identify DNA relatives using accessible databases
6.2.9	Ability to conduct and evaluate genealogical research and construct family trees
6.2.10	Ability to identify common ancestors and union couples using genetic genealogy
6.2.11	Knowledge of DNA probabilities and their use to hypothesize familial relationships
6.2.12	Ability to conduct descendancy research and to build family trees forward to identify deceased and living persons
6.2.13	Ability to use genetic genealogy tools

(*Continued*)

TABLE 8.2 (Continued)

Section	Description of Competency
6.2.14	Ability to formulate and test hypotheses, and to make recommendations to advance an investigation
6.2.15	Ability to communicate research and recommendations

8.6.4 Transparency, Accountability & Public Trust

Professional genealogists, including AIGGs, must maintain high standards of professional integrity. This includes maintaining transparency and accountability and engaging in conduct that will build public trust in the field.

The BCG Code of Ethics sets out a number of professional obligations, including the duty to conduct oneself in a manner that is in the best interests of the genealogical profession and genealogical scholarship, to expose genealogical fraud, to refrain from practices such as plagiarism or passing off the work of another genealogist, as well as a duty to reveal conflicts of interest that could compromise one's work as well as duties of candor [6].

The IGGAB Code of Ethics sets out several duties of transparency and accountability for AIGGs, including a duty to identify themselves and explain their role to any relatives or Reference Test Takers, as set out in section 5.1(a), as well as the methods and techniques used, set out in section 5.1(b). AIGGs must be honest and candid about their level of skill and knowledge, as per sections 5.1(c) and (d), and must disclose any conflicts of interest, as per section 6.1(a) [2].

Other professional duties of AIGGs include integrity and objectivity, set out in section 6.1(a) [2]. AIGGs must refrain from certain activities that could discredit the profession or the field of FIGG, as per section 6.1(c), including by not soliciting cases being worked by another practitioner, as per section 6.1(b) [2]. The IGGAB Code of Ethics also imposes professional obligations of honesty and candor, as set out in section 6.1(e), civility and courtesy as per section 6.1(d), and the duty to collaborate with other professionals in an ethical and efficient manner as per section 6.1(f) [2].

The IGGAB Professional Standards require AIGGs to self-report any breaches of the professional standards, set out in section 7(a), as well as breaches that have been, to the best of their knowledge, committed by another AIGG, as per section 7(b) [3]. Sections 5.2 (a) and (b) of the IGGAB Code of Ethics also require AIGGs to promptly address any error and respond promptly to the IGGAB [2]. AIGGs also have a duty to respond promptly to all communications from the IGGAB, as set out in section 7(c), and to promptly provide any materials requested of them by the IGGAB as per section 7(d), so that the board may ensure compliance with its standards [3].

8.7 SUMMARY

The practice of FIGG has raised several ethical issues in addition to those posed by genealogical research, genetic genealogy, and forensic DNA. Ethical norms and professional standards have been developed to address these issues, mainly through various self-regulating professional bodies in the field. Many of these norms have been—and are being—worked out on an *ad hoc* basis, often as a result of ethical controversies that have arisen in the field.

Several bodies have issued ethical and professional guidelines to govern the conduct of FIGG. The IGGAB has issued a Code of Ethics and a set of Professional Standards, developed with the input of members of the public, genetic genealogists, and practitioners of FIGG [1–3]. These standards set out some of the key ethical duties imposed on AIGGs, including respecting the individual autonomy of DNA test takers, including their bodily integrity and their right to control the sharing and use of their information, as well as the need for informed consent for Reference Test Takers. The NTVIC has also issued a set of guidelines for the establishment of FIGG programs, which include guidelines for practitioners of FIGG [9, 10].

The IGGAB standards also include professional obligations of honesty and candor, high standards for knowledge, competence, and quality of work, including knowledge of forensic DNA, and professional ethics. The ethics of FIGG should remain focused on key goals that the field strives to accomplish: respect for the autonomy, privacy, and dignity of individuals who find themselves within the scope of a FIGG investigation, and building public trust in this emerging field of forensics.

REFERENCES

1. Investigative Genetic Genealogy Accreditation Board (IGGAB) [Internet]. About Us; 2024 [cited 2025 Jan 12]. Available from: https://www.iggab.org/
2. Investigative Genetic Genealogy Accreditation Board (IGGAB). Investigative Genetic Genealogy – Code of Professional Ethics. *IGGAB C001*. 2025 Apr. https://www.iggab.org/uploads/1/4/2/9/142901820/c001_iggab_code_of_professional_ethics_publication.pdf.

3. Investigative Genetic Genealogy Accreditation Board (IGGAB). Investigative Genetic Genealogy – Professional Standards and Accreditation Requirements. *IGGAB S001*. 2025 Apr. Supp. https://www.iggab.org/uploads/1/4/2/9/142901820/s001_investigative_genetic_genealogy_professional_standards_and_accreditation_requirements_v2.pdf.
4. The Genetic Genealogy Standards Committee [Internet]. *Genetic genealogy standards*; 2015 Jan 10. Available from: https://thegeneticgenealogist.com/2015/01/10/announcing-genetic-genealogy-standards/
5. Board for Certification of Genealogists. *Genealogy standards, 2nd edition, revised*. Nashville, Ancestry; 2021.
6. Board for Certification of Genealogists. Code of ethics, Appendix A. In: *Genealogy standards*, 2nd ed. Turner Publishing; 2019. p. 49–53.
7. Board for Certification of Genealogists [Internet]. *Certified Genetic GenealogistSM Application Instructions*; 2024 Mar 1 [cited 2025 Jan 12]. Available from: https://www.bcgcertification.org/become-a-certified-genealogist#application-guide
8. Council for the Advancement of Forensic Genealogy [Internet]. Standards of practice and conduct. 2025 [cited 2025 Jan 12]. Available from: https://www.forensicgenealogists.org/about-us/standards
9. Wickenheiser RA, Naugle J, Hoey B, Nowlin R, Kumar SA, Minton A, Allen L, Glynn C. National technology validation and implementation collaborative (NTVIC) policies and procedures for forensic investigative genetic genealogy (FIGG). *Forensic Science International: Synergy* 2023 Jan 1;6:100316. DOI: https://doi.org/10.1016/j.fsisyn.2023.100316
10. Wickenheiser RA, Naugle J, Hoey B, Nowlin R, Kumar SA, Kubinski MA, Glynn C, Valerio R, Allen L, Stoiloff S, Kochanski J. National technology validation and implementation collaborative (NTVIC): guidelines for establishing forensic investigative genetic genealogy (FIGG) programs. *Forensic Science International: Synergy.* 2023;7. DOI: https://doi.org/10.1016/j.fsisyn.2023.100446
11. Md. Code Ann., Crim. Proc. Title 17 (2023).
12. Maryland Department of Health [Internet]. Office of Health Care Quality; c2025 [cited 2025 Jan 12]. Available from: https://health.maryland.gov/ohcq/Pages/home.aspx
13. Eyal N. Informed Consent. In: Miller A, Schaber P, editors. *The Routledge handbook of the ethics of consent*. Oxon/New York: Routledge; 2018. p. 272–284.
14. Logan v Ontario (Solicitor General), 2022 HRTO 1004 (CanLII). Available from: https://canlii.ca/t/jrjkm

15. Osmond RV, 2009 BCSC 550 (CanLII). Available from: https://canlii.ca/t/23hpf
16. Russel JG. Ethical Underpinnings of Genetic Genealogy. In Wayne DP, editor. *Advanced genetic genealogy: techniques and case studies*. Cushing, TX: Wayne Research; 2019. p. 303–324.
17. Careau J, Larmuseau MH, Drumsta R, Whitley R. I'm trying to figure out who the hell I am: Examining the psychosocial and mental health experience of individuals learning "not parent expected" news from a direct-to-consumer DNA ancestry test. *BMC Psychiatry.* 2025 [cited 2025 Jan 12]; 25(9). DOI: https://doi.org/10.1186/s12888-024-06380-0.
18. Guerrini CJ, Robinson JO, Bloss CC, Bash Brooks W, Fullerton SM, Kirkpatrick B, et al. Family secrets: experiences and outcomes of participating in direct-to-consumer genetic relative-finder services. *American Journal of Human Genetics* 2022 [cited 2025 Jan 12];109(3):486–497. DOI: https://doi.org/10.1016/j.ajhg.2022.01.013.
19. Anderson KG. How well does paternity confidence match actual paternity? Evidence from worldwide nonpaternity rates. *Current Anthropology* 2006;47(3):513–520.
20. Jiminez C. My whole life, I embraced my Latina identity. *Then I took a DNA test that changed everything*. Buzzfeed [Internet]. 2024 Dec 27 [cited 2025 Jan 12]. Available from: https://www.buzzfeed.com/cubajimenez/dna-test-results-surprise-ancestry
21. Yiaueki S. I was raised as a Native American. *Then a DNA test rocked my identity*. The Guardian [Internet]. 2018 Nov 15 [cited 2025 Jan 12]. https://www.theguardian.com/commentisfree/2018/nov/15/raised-native-american-dna-test-father-lied-heritage
22. Ross J. *Woman's DNA test sent her grandma to prison [video on the Internet]*. You Tube: Professional Genealogist Reacts; 2024 Dec 23 [cited 2025 Jan 12]. Available from: https://www.youtube.com/watch?v=fOpNPJ8D6Ps
23. Galgano T. She took a DNA test for fun. *Police used it to charge her grandmother with murder in a cold case*. CNN [Internet]. 2024 Dec 12 [cited 2025 Jan 12]. https://www.cnn.com/2024/12/12/us/baby-garnet-cold-case-michigan/index.html
24. Miller FG, Wertheimer A, editors. *The ethics of consent*. New York: Oxford University Press; 2010.
25. Sreenivasan G. Does informed consent to research require comprehension? *Lancet*. 2003;362: 2016–2018.
26. Wrigley A. Consent for Others. In Miller A, Schaber P, editors. *The Routledge handbook of the ethics of consent*. Oxon/New York: Routledge; 2018. p. 322–334.

27. Family TreeDNA. Privacy statement [Internet]. 2024 Aug 6. Last modified 2024 Aug 6 [cited 2025 Jan 12]. Available from: https://www.familytreedna.com/legal/privacy-statement
28. FamilyTreeDNA. Law enforcement guide [Internet]. 2024 Apr 25. Last modified 2024 Apr 11 [cited 2025 Jan 12]. Available from: https://www.familytreedna.com/legal/law-enforcement-guide
29. GEDmatch. Privacy policy [Internet]. 2024 Nov 22 [cited 2025 Jan 12]. Available from: https://www.gedmatch.com/privacy-policy-november-22-2024/
30. GEDmatch Pro. Terms of use [Internet]. 2024 Oct 28 [cited 2025 Jan 12]. Available from: https://pro.gedmatch.com/terms-use-oct-2024
31. GEDmatch. Terms of service [Internet]. 2024 Nov 22 [cited 2025 Jan 12]. Available from: https://www.gedmatch.com/terms-of-service-november-22-2024/
32. DNA Justice Foundation. Terms of service [Internet]. c2024 [cited 2025 Jan 12]. Available from: https://www.dnajustice.org/tos
33. DNA Justice Foundation. Privacy notice [Internet]. 2022 Oct 24 [cited 2025 Jan 12]. Available from: https://www.dnajustice.org/privacy
34. Othram. DNASolves® Connect [Internet]. 2025 [cited 2025 Feb 17]. Available from: https://connect.dnasolves.com/
35. GEDmatch. GEDmatch.com terms of service and privacy policy [Internet]. 2019 May 18 [cited 2025 Jan 12]. Available from: https://web.archive.org/web/20190612170606/https://www.gedmatch.com/tos.htm
36. Press M. From our founder, Margaret Press [Internet]. 2023 Aug 18 [cited 2025 Jan 12]. Available from: https://dnadoeproject.org/statement-from-Margaret-Press/
37. Russell J. Truth. 2023 Sep 17 [cited 2025 Jan 12]. In: The Legal Genealogist [Internet]. Available from: https://www.legalgenealogist.com/2023/09/17/truth/
38. Kumar SA. Notice regarding investigations into FIGG practitioners circumventing GEDmatch settings and violating terms of service, and actions taken. *GEDmatch News [Internet]*. 2023 Nov 13 [cited 2025 Jan 12]. Available from: https://verogen.com/notice-regarding-investigations-into-figg-practitioners-circumventing-gedmatch-settings-and-violating-terms-of-service-and-actions-taken/
39. Larkin LP. Noble cause, no rules.2023 Jun 27 [cited 2025 Jan 12]. In: The DNA Geek [Internet]. Available from: https://thednageek.com/noble-cause-no-rules/
40. Vogen M. Riverside County DA's Office, Sheriff's Department, and FBI leverage Othram's genetic testing platform to identify 1996 homicide victim. *DNASolves.com [Internet]*. 2023 Jun 27 [cited 2025 Jan 12]. Available from: https://dnasolves.com/articles/riverside-county-juana-rosas-zagal/

41. MyHeritage. Privacy policy [Internet]. 2024 Sep 11 [cited 2025 Jan 12]. Available from: https://www.myheritage.com/privacy-policy
42. United States Department of Justice. United States Department of Justice interim policy: forensic genetic genealogical DNA analysis and searching [Internet]. 2019. Available from: https://www.justice.gov/olp/page/file/1204386/download
43. UTAH CODE §53-10—403.7.
44. H.B. 602, 2021 Rg. Sess. (Mt. 2021).
45. Dowdeswell T. Forensic genetic genealogy project v. JULY2024. 2024 Sep 16 [cited 2025 Jan 12];1. Available from: https://data.mendeley.com/datasets/z5khb77xyd/1
46. Vaughan A. Home DNA-testing firm will let users block FBI access to their data. New Scientist [Internet]. 2019 Mar 13 [cited 2025 Jan 12]. Available from: https://www.newscientist.com/article/2196433-home-dna-testing-firm-will-let-users-block-fbi-access-to-their-data/
47. Federal Bureau of Investigation. *Crime Data Explorer [Internet]*. 2024 [cited 2025 Jan 12]. Available from: https://cde.ucr.cjis.gov/LATEST/webapp/#/pages/explorer/crime/crime-trend
48. Federal Bureau of Investigation. Crime in the United States 2013: Rape [Internet]. 2013 [cited 2025 Jan 12]. Available from: https://ucr.fbi.gov/crime-in-the-u.s/2013/crime-in-the-u.s.-2013/violent-crime/rape
49. Christman MA, Piquero AR. New and better crime data for the nation. *U.S. Department of Justice [Internet]*. 2022 Oct 5 [cited 2025 Jan 12]. Available from: https://www.justice.gov/opa/blog/new-and-better-crime-data-nation
50. Criminal Code. R.S.C. 1985, c. C-46.
51. D.C. CODE. §§ 22-3001 – 22-3024.
52. Aldhous P. The arrest of a teen on an assault charge has sparked new privacy fears about DNA sleuthing. *BuzzFeed News [Internet]*. 2019 May 14 [cited 2025 Jan 12]. Available from: https://www.buzzfeednews.com/article/peteraldhous/genetic-genealogy-parabon-gedmatch-assault
53. Miller SJ A Centerville teen faces aggravated assault allegations for attacking an elderly woman in a locked church. *Salt Lake Tribune*. May 24, 2019. https://www.sltrib.com/news/2019/05/24/centerville-teen-faces/
54. GEDmatch. GEDmatch.com terms of service and privacy policy [Internet]. 2018 May 20 [cited 2025 Jan 12]. Available from: https://web.archive.org/web/20180527131537/https://www.gedmatch.com/tos.htm

55. Rogers CA. Forensic revolution: IGG investigative genetic genealogy. *Authorlink [Internet]*. 2024 Jun 7 [cited 2025 Jan 12]. Available from: https://authorlink.com/publishers-market/a-forensic-revolution-igg-investigative-genetic-genealogy/
56. Aldhous P. This genealogy database helped solve dozens of crimes. But its new privacy rules will restrict access by cops. *BuzzFeed News [Internet]*. 2019 May 19 [cited 2025 Jan 12]. Available from: https://www.buzzfeednews.com/article/peteraldhous/this-genealogy-database-helped-solve-dozens-of-crimes-but
57. GEDmatch. GEDmatch.com terms of service and privacy policy [Internet]. 2021 Dec 30 [cited 2025 Jan 12]. Available from: https://www.gedmatch.com/terms-of-service-privacy-policy/
58. Saey H. Genealogy companies could struggle to keep clients' data from police. *Science News [Internet]*. 2019 Jun 10 [cited 2025 Jan 12]. Available from: https://www.sciencenews.org/article/forensic-genetic-genealogy-companies-police-privacy
59. Hill K, Murphy H. Your DNA profile is private? A Florida judge just said otherwise. *New York Times [Internet]*. 2019 Nov 5 [cited 2025 Jan 12]. Available from: https://www.nytimes.com/2019/11/05/business/dna-database-search-warrant.html#
60. Ancestry 2019 Transparency Report [Internet] [cited 2025 February 14]. Available from: https://www.ancestry.com/cs/transparency-2019
61. St. John P. DNA genealogical databases are a gold mine for police, but with few rules and little transparency. *Los Angeles Times [Internet]*. 2019 Nov 24 [cited 2025 Jan 12]. Available from: https://www.latimes.com/california/story/2019-11-24/law-enforcement-dna-crime-cases-privacy
62. Aldhous P. A court tried to force Ancestry.com to open up its DNA database to police. The company said no. *BuzzFeed News [Internet]*. 2020 Feb 3 [cited 2025 Feb 14]. Available from: https://www.buzzfeednews.com/article/peteraldhous/ancestry-dna-database-search-warrant
63. Associated Press. Green River Killer's youngest known victim identified through genetic genealogy. *Oregon Live [Internet]*. 2021 Jan 26 [cited 2025 Jan 12]. https://www.oregonlive.com/crime/2021/01/green-river-killers-youngest-known-victim-identified-through-genetic-genealogy.html
64. Casarez J, Griggs B. This 'CSI: Miami' actress was threatened by a stalker for 12 years. The FBI caught him after he left his DNA on a fast-food straw. *CNN [Internet]*. 2022 Sep 26 [cited 2025 Jan 12]. Available from: https://www.cnn.com/2022/09/09/entertainment/eva-larue-stalker-sentenced/index.html

65. Oliver B. Brazos County District Court jury convicts then issues a life sentence for burglary and attempted sexual assault. *WTAW* [Internet]. 2022 Aug 29 [cited 2025 Jan 12]. Available from: https://wtaw.com/brazos-county-district-court-jury-convicts-then-issues-a-life-sentence-for-burglary-and-attempted-sexual-assault/
66. Williams V. Texas. No 06-2200142 CR (CA 2023). Available from: https://scholar.google.ca/scholar_case?case=8595351330375906077&q=Christopher+Williams&hl=en&as_sdt=4,238,240,241&as_ylo=2021
67. TEX. PENAL CODE ANN. § 22.011, § 30.02 (2020).
68. TEX. CODE CRIM. PROC. ANN. § 62.001 (2020).
69. Order M, Miller W. College Station police arrest man tied to several assaults near Holleman. *KBTX News [Internet]*. 2018 Dec 11 [cited 2025 Jan 12]. Available from: https://www.kbtx.com/content/news/College-Station-police-arrest-man-tied-to-several-assaults-near-Holleman-502482482.html
70. City of Riverside. Non-verbal found adult finally identified. *City of Riverside [Internet]*. 2019 Jul 12 [cited 2025 Jan 12]. Available from: https://www.riversideca.gov/press/non-verbal-found-adult-finally-identified
71. DeForest M. Benjamin Holmes found guilty in 2001 murder of UCF student. *ClickOrlando.com [Internet]*. 2023 Feb 7 [cited 2025 Jan 12]. Available from: https://www.clickorlando.com/news/local/2023/02/08/benjamin-holmes-found-guilty-in-2001-murder-of-ucf-student/
72. Shuppe J. 'They lied to us': Mom says police deceived her to get her DNA and charge her son with murder. *NBC News*, February 22, 2020. https://www.nbcnews.com/news/us-news/they-lied-us-mom-says-police-deceived-her-get-her-n1140696
73. Perez N, Torres A. Ancestry customer says he is glad his DNA on GEDmatch led to unknown cousin's arrest for murder. *Local 10 News [Internet]*. 2019 Aug 14 [cited 2025 Jan 12]. Available from: https://www.local10.com/news/2019/08/15/ancestry-customer-says-he-is-glad-his-dna-on-gedmatch-led-to-unknown-cousins-arrest-for-murder/
74. State v Athan, 158 P.3d 27 (2007).
75. D'Amico c R, 2019 QCCA 77 (CanLII). Available from: https://canlii.ca/t/hx4t6

CHAPTER 9

Artificial Intelligence and Genealogy

Nicoladie Tam

CONTENTS

 DOI: 10.4324/9781032714998-9

9.1 INTRODUCTION

Our uniqueness arises from genetic variations that manifest as distinct phenotypic traits. Environmental factors, including epigenetics, also impact gene expression, resulting in diverse phenotypes (or characteristics) even among individuals with the same genotype (or gene sequences), such as identical twins who vary due to environmental influences. The interactions between environments and genes create variations among us that contribute to our uniqueness. Without these variations, we would resemble factory-produced robots, distinguishable only by their serial numbers.

9.1.1 An Overview of the Quantitative Assessment of Genetic Data

The methodology for identifying an individual's or population's ancestral lineage or inherited traits varies based on project objectives, particularly in forensic and genealogical studies. This overview discusses methods for quantitatively assessing genetic information and includes case examples to illustrate the outcomes of the applied analytical techniques. The theoretical principles of these methods are presented to facilitate an objective evaluation of the results. Additionally, the advantages and disadvantages of each technique are examined to support informed decision-making. Data availability, technology options, and performance outcomes can influence the selection of analytical tools and the necessity of definitive conclusions for the project's goals.

9.1.1.1 Rationale for Choosing Specific Biometric Markers

Evolution offers evidence for tracing generations through inherited patterns in gene sequences. These variations assist in identifying genealogical lineages and reveal the divergent branches that enhance our understanding of identity and ancestry. The gene sequences provide quantitative evidence of inherited patterns transmitted from prior generations.

9.1.1.2 The Use of Genetic Codes as Identifying Markers

Analyzing gene patterns can reveal previously unknown genealogical lineages based on inherited traits in family trees. Constructing a genealogical tree involves connecting members according to the likelihood of these patterns being passed down. This process infers a descendant's inheritance from the similarities and differences in gene sequences, organizing the tree by positioning members based on the most probable inherited traits. The gene code pattern can be a fingerprint identifier for matching an individual's identity in forensic determinations.

9.1.1.3 The Human Genome as an Identifying Blueprint

Deoxyribonucleic acid (DNA) encodes genetic information through its molecular sequence found within the genes of all living organisms, from bacteria to humans. DNA provides the genetic instructions for protein synthesis, cell replication, and repair. It also influences the developmental timing of plants and animals, affecting their resilience and survival.

DNA serves as the genetic code found in a cell's chromosomes. The complete set of chromosomes is referred to as the genome. The human genome consists of 6 billion nucleotides organized into 23 pairs of chromosomes. The number of chromosomes varies across species; for instance, some butterflies possess over 400, whereas certain single-celled protozoa have 1600.

9.1.1.4 The Complete DNA Codes in the Human Genome

Nucleotides form long polymer chains that create the backbone of a double-stranded helix composed of deoxyribose and phosphate groups. Strong covalent bonds link the nucleotides, while weaker hydrogen bonds connect the strands, resulting in a double helix with complementary base pairs. Each strand pairs A with T and G with C at corresponding loci on the chromosomes. The complementary base pairs enable one strand to predict its counterpart, resulting in three billion unique nucleotide codes that reflect individual similarities and differences.

9.1.1.5 The Complexity of Genetic Codes

Analyzing three billion DNA codes in the human genome is a daunting task, whether the goals are to identify individuals for forensics, trace ancestral origins for genealogy, or connect genetic factors associated with disorders for medical research. Examining the similarities and differences can provide insights into our genetic heritage.

9.1.2 Genome-Wide Association Studies (GWAS)

Although the human genome sample size is substantial, various high-throughput techniques allow for the systematic analysis of DNA sequence variations across populations in genome-wide association studies (GWAS). These techniques identify genomic variants statistically linked to specific traits or the risk of a disease, providing insights into the genome's coding and non-coding regions.[1, 2]

9.1.2.1 The Use of Gene Variants to Minimize Size for Comparison

Human nucleotide sequence variations range from 0.1% to 0.4%. While DNA sequences share over 99% similarity, the combinatorial variations remain significant for the comparative analysis of three billion-letter codes. Fortunately, most variations consist of single-nucleotide substitutions or repetitive patterns. Comparing short DNA segments with variations in a population is sufficient for tracing or identification. These short sequences allow for comparison without needing to sequence the entire genome.

9.1.3 Distinct Short DNA Sequences as Biometric Identifiers for Tracking

Even though each individual's unique genome can serve as a biometric identifier that distinguishes one person from another, it is commonly called a biomarker when a shorter sequence is used. Variations in these gene biomarkers can verify individual similarities and differences, including susceptibility to disorders, behavioral outcomes, and forensic tracing.

9.1.3.1 The Application of Gene Variants in Medical and Genealogical Analyses

These gene variants clearly distinguish individuals, irrespective of their functions or the traits they influence. They serve as objective biometric markers for tracing ancestry, locating relatives, confirming parenthood, verifying an individual's identity, unveiling genetic predispositions to disorders, and providing insights into heritage.

9.1.3.2 The DNA Sequences for Analysis

These DNA sequences are the foundation for all genetic coding information, including coding and non-coding regions in the chromosomes. This information can be used for forensic or genealogical analysis and to identify genetic links to medical disorders. While various forms of data, such as historical records and personal accounts, can provide valuable insights for forensic and genealogical tracing, this chapter focuses on leveraging DNA data for these purposes.

9.1.4 The Role of Non-Coding DNA in Forensic Analysis

Not all gene sequences encode or regulate gene expression; some are non-coding and unexpressed. Coding regions are essential for studying effects on growth, development, and medical disorders.[3] Since non-coding sequences are not expressed, they do not typically influence trait outcomes, which can affect analysis. For these reasons, non-coding regions are commonly used for forensic or genealogical analysis.

9.1.4.1 Effects of Non-Coding DNA on Disorders

Although non-coding regions are not transcribed, they are associated with specific phenotypic variants identified by the GWAS project. Single-nucleotide polymorphisms (SNPs) in non-coding DNA can influence intron activity and regulatory elements, such as promoters and enhancers. These variants may impact DNA methylation, histone modifications, transcription factor affinity, alternative splicing, and mRNA stability. These factors suggest that they can alter traits even though they are not directly transcribed.[4] In forensic analysis, it was once assumed that using non-coding regions would not affect traits, but emerging evidence may not support this assumption.

9.1.4.2 Amplification of DNA Sequence for Analysis

The polymerase chain reaction (PCR) amplification method is widely used to replicate a specific DNA sequence for analysis. Sequencing a shorter segment is simpler than sequencing the entire genome. Several other amplification and sequencing techniques are available for extracting a target DNA sequence for analysis.

9.1.5 Short Sequence Variants as Biomarkers for Identification

To reduce the sequence size for analysis, one can select a specific short sequence for comparison, provided that the sequence distinguishes differences among individuals as identifiable markers. These short sequences can serve as biomarkers for assessing the likelihood of inheritance within a lineage, thereby narrowing the scope of comparison. When these short nucleotide segments provide sufficient data to trace inherited traits, they can yield enough information to create a genetic profile for forensic analysis.

9.1.5.1 Improve Matching Accuracy with Multiple Biomarkers

False positives inevitably occur in any testing due to errors, regardless of accuracy. Utilizing short sequence biomarkers can match genetic traits more efficiently than analyzing the entire genome; however, false positives persist. Nevertheless, the error rate can be significantly reduced by employing multiple biomarker sequences from different loci (locations on a gene) instead of relying on just one.

For example, with a 10% false positive error rate (1 in 10) per sequence tested, two sequences result in a 1% error chance (10% × 10% = 1 in 100). Testing three sequences reduces this to 0.1% (10% × 10% × 10% = 1 in 1000). With 13 sequences, the false positive rate drops to 0.113% or 1 in a trillion. Thus, most forensic analyses use multiple biomarker sequences at various loci to determine matches.

9.2 AN OVERVIEW OF TRADITIONAL STATISTICAL INFERENCE

Identifying genealogical lineage relies on methods that evaluate the probability of inheriting DNA sequences from ancestors or matching individuals through genetic profiles instead of depending on fingerprint analysis. Traditional approaches use statistical inferences to assess the likelihood of trait transmission, which aids in constructing genealogical trees or matching DNA biomarker sequences for forensic analyses.

9.2.1 False Positives and Statistical Confidence

These methodologies rely on the statistical probability of matching a target population or gene sample. However, the possibility of error always exists. To minimize the risk of false positives, statistical methods lower error rates by increasing the confidence intervals in the analysis. A conclusion can only be drawn when there is sufficient statistical probability to establish a genealogical tree or evaluate the likelihood of matching an individual to a target sample.

9.2.2 Dependency On a Specific Database

However, comparing genetic data requires access to a database of DNA sequences from populations potentially related to the individual. The location on the tree can be identified if the matching sequences can be traced back to the genetic profiles already present in the database. Tracing the location on the tree becomes increasingly difficult when related profiles are absent from the database.

9.3 AN OVERVIEW OF ARTIFICIAL INTELLIGENCE IN GENEALOGY

This overview of using artificial intelligence (AI) in genealogy summarizes the capabilities of AI models. It discusses the advantages and limitations of this technology before exploring the differences among various AI models that impact genealogical analysis. Subsequent sections will present a more detailed discussion, featuring a case history and examples to illustrate the principles used in AI, offering objective criteria for evaluating the performance of different AI models in genealogy.

9.3.1 The Alternative AI Approach Beyond Statistical Inferences

In contrast to the traditional standard approach, artificial intelligence techniques can analyze and generalize genetic information more effectively than conventional methods. They provide potential solutions that uncover new insights into genealogical inheritance and bridge data gaps through generalizations. Natural language processing AI models can automate the categorization of written genealogical records[5] while offering human-like interactions to address inquiries related to genealogy by generalizing genetic profile information from the training datasets.

9.3.2 Addressing the Data Gaps

AI effectively addresses issues related to missing data. Current AI technology employs machine learning (ML) to integrate extensive datasets and train systems to establish connections between data points. Once trained, these connections generate outputs from input queries, even when specific data points are absent from the training set. The system can interpolate missing information by leveraging the connectivity among relevant data. In forensic genealogy, if the database lacks genetic profile information,

AI can estimate likely tree locations, which traditional algorithms may struggle to accomplish.

9.3.3 Generalizing Genealogical Patterns

Other applications of AI models in genealogy assess relationships without depending on traditional statistical methods. Rather than explicitly programming analytical techniques into computer software, AI models generate results for input queries by generalizing patterns learned from the interactions between inputs and outputs in the training dataset.

9.3.3.1 Responding to Genealogical Inquiries Using Natural Language Processing

For instance, AI can generate natural responses to genealogical inquiries using a large language model (LLM), such as ChatGPT. It can automate user requests and provide answers without human involvement. Generative AI can create interactive conversations with users to address questions based on training with a vast dataset.

9.3.4 Cataloging Natural Language Genealogical Records

Applying AI in genealogy can extract information from historical records, including newspaper articles and oral histories, and organize it into databases. This cataloging process is often labor-intensive; automating it would minimize the need for human involvement and speed up information retrieval. Users can benefit from organized genealogical information for future queries and can pose questions to receive more efficient answers.

9.3.5 Machine Learning (ML) for Learning from Training Datasets

An AI system develops internal models by learning input-output relationships through ML without relying on human-provided solutions or algorithms. ML automates this process by applying learning rules to establish connections, including direct and higher-order relationships among relevant factors. The system utilizes a training dataset to improve its internal representations through trial and error, using an error minimization methodology to achieve optimal outcomes. It requires billions of examples in the dataset and billions of iterations in the trial-and-error process to converge on a solution, effectively linking input queries to system outputs.

9.3.5.1 Biases in Training Databases

The model may unintentionally introduce bias by optimizing internal representations based on a particular training dataset. While large datasets are essential, biased outcomes can arise if counterexamples are excluded. Due to the limited genealogical databases, conclusions drawn from specific datasets might be flawed. Furthermore, outliers are often misrepresented during the generalization process.

9.3.5.2 Limitations of Artificial Intelligence

The process by which the AI system reaches its conclusions is often unclear. It generates outputs by learning correlations from extensive datasets. Consequently, AI models excel at generalizing key data to produce results. They do not apply logical reasoning or specific rationale to deduce answers, such as using physiological or genetic knowledge of inheritance or employing statistical principles to infer outcomes. Therefore, AI outputs typically lack explanations unless explainable AI (XAI) is used to clarify the reasons based on the analysis of statistical input-output patterns. Without logical inferences and reasoning rationales, the AI system's conclusions may be plausible yet lack rationality.

9.4 GENEALOGICAL ANALYSIS METHODOLOGIES

Methods for genealogical analysis rely on the interactions between genotypes and phenotypes. Variations in DNA sequences can lead to differences in genotypic and phenotypic traits. These genotypes influence our physical characteristics, closely linking them to phenotypic qualities. Such variations emerge through the mixing and matching processes during DNA replication and recombination. Analyzing these sequences can uncover the genetic influences on inherited traits, which may offer survival advantages or disadvantages. As we examine probabilities on a genealogical tree, these variations can affect lineage branching and extinction.

9.4.1 Analysis of DNA Patterns

As mentioned earlier in the introduction, genetic codes are represented by the four-letter symbols A, C, G, and T. These sequences correspond to their respective chromosome pairs, with A pairing with T and C pairing with G. The analytical methodology focuses on the patterns of similarities and differences among a series of DNA sequences.

9.4.1.1 The Degree of Similarity and Difference

In genealogical tracing, the degree of similarity can indicate a level of relatedness. Identifying lineage becomes straightforward when comparing identical and non-identical codes in a DNA sequence as measures of relatedness. However, exact and non-identical sequences do not necessarily correlate with similarities or dissimilarities or imply inheritance for the reasons explained below.

9.4.1.2 The Difference Defined by a Single-Nucleotide Sequence

When a DNA sequence varies by one or two nucleotides, should it be considered similar or different compared to a group? Classification is relative to the population with which it is compared and is influenced by levels of variability. Similarities or differences can change depending on the subpopulation used for comparisons. Assigning an individual to a lineage tree depends on their similarities and variations compared to other groups. The analysis involves more than just genetic similarities and differences; it requires additional methods to trace lineages based on how descendants inherit genes and survive.

9.4.2 Principles of Inheritance

Various factors influence inheritance patterns in sexual reproduction. This reproductive method accelerates the evolutionary rate by mixing recombinant DNA more effectively than random mutations, thereby enhancing survival. Consequently, the recombinant DNA sequences differ from those of either parent, while descendants inherit some traits but not all. Understanding these processes could assist in identifying suitable analytical tools for genealogical analysis.

9.4.2.1 Inheritance via Asexual Reproduction

In asexual reproduction through cloning from a single parent, the offspring's DNA is identical to that of the parent, as they are exact replicas of the organism. When cloned, the progeny of asexual reproduction inherit 100% of the parent's traits, assuming there are no mutations or replication errors during reproduction. The process of inheritance in asexual reproduction is straightforward and certain. However, inheritance in sexual reproduction, which involves recombinant DNA, is more complex and probabilistic rather than certain.

9.4.2.2 Inheritance via Sexual Reproduction

Sexual reproduction combines DNA from both parents, producing offspring that inherit half of their traits from each. It accelerates evolution by mixing and matching in a trial-and-error recombinant reproductive process. Unlike asexual reproduction, the inheritance patterns in sexual reproduction are not definitive; they are probabilistic and depend on numerous factors that affect genealogical analysis.

9.4.2.3 Single-Stranded Chromosomes in Haploid Cells

During sexual reproduction, meiosis is a type of cell division that separates chromosomes into unpaired forms, resulting in haploid gametes, specifically sperm and egg cells. Each gamete contains one copy of each chromosome, made up of a single DNA strand. The chromosomes remain unpaired until fertilization, when the sperm and egg fuse together.

9.4.3 Double-Stranded Chromosomes in Diploid Cells

During fertilization, a sperm cell inserts its nucleus into an egg, recombining chromosomes into paired strands. One unpaired chromosome is inherited from the mother and another from the father, resulting in a diploid cell. The fertilization process recombines unpaired DNA strands, giving the embryo half of the traits from each parent, while the chance of inheriting a specific strand remains probabilistic.

9.4.3.1 The Randomness of Producing a Male or Female Offspring

Sexual reproduction demonstrates the randomness of inheriting specific sex chromosomes. Parents can produce offspring with either XX chromosomes (indicating females) or XY chromosomes (indicating males) (see Figure 9.1). This process occurs by chance during fertilization. The likelihood of having a boy or a girl is a probabilistic phenomenon rather than an absolute certainty.

9.4.3.2 The Probability of Having a Girl or a Boy

The probability of inheriting a particular sex chromosome, such as the XX chromosome, is 50% (2 out of 4) for girls and 50% (2 out of 4) for

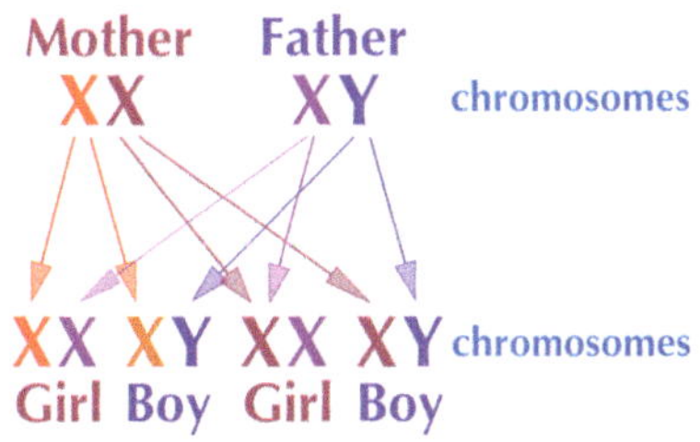

FIGURE 9.1 The diagram illustrates the recombinant X and Y chromosomes, showing a 50% (2 out of 4) probability of inheriting either an XX chromosome for a girl or an XY chromosome for a boy.

FIGURE 9.2 The inheritance table illustrates the recombinant XX and XY chromosomes from the mother and the father, similar to the inheritance shown in Figure 9.1.

boys (see Figure 9.2), assuming the fertilization process is not manipulated. It indicates the chance of having a girl or a boy is 50/50. This example illustrates the random chance involved in inheriting traits through sexual reproduction.

9.4.4 The Influence of Survival Rates on Prevalence Statistics

Due to survival rates, population prevalence statistics can overestimate or underestimate inheritance probabilities. For example, females tend to live longer, leading to a male-to-female ratio of 49% to 51%. Prevalence data may misrepresent the frequency of gene variants based on survival rates: a male-linked variant shows a bias toward 49%, while a female-linked variant exhibits a bias toward 51%. This example illustrates how survival hazard functions influence age-specific incidence statistics. Therefore, prevalence statistics may not accurately reflect occurrences of gene variants unless age-specific survival rates are considered.

9.4.4.1 The Hazard Function of Survival Rate for Each Age Group

The hazard function represents the cumulative distribution until death occurs. If a gene variant leads to early childhood deaths, those cases will not be counted if the deaths occurred before the population survey. Furthermore, if parents do not have children, the chances of passing down genes are nonexistent, not due to death but rather a lack of reproduction. The prevalent statistics can be adjusted for the survival rate within each age group for a more accurate representation.

9.4.4.2 Factors Influencing Parental Inheritance

Although all offspring's genes can be traced back to their parents, one-half of the parents' chromosomes are absent in the offspring. When a gene variant is used to track inheritance, that variant may or may not be present in the descendant's genes. The offspring might inherit the normal gene from the heterozygous carrier parent and receive a regular copy instead of the variant.

Forensic and genealogical tracing requires an analysis of how genes are passed down from each parent, which is crucial for drawing reliable conclusions. The probability of passing a gene variant to offspring depends on whether one or both parents carry it and whether they are heterozygous or homozygous carriers.

9.4.5 Estimating the Likelihood of Inheriting a Gene Variant

Unlike asexual reproduction, the probability of passing this variant to the next generation is generally considered to be 50% for an offspring inheriting one chromosome from each parent. This scenario assumes that one parent is normal. In contrast, the other parent carries the variant and is heterozygous for that gene variant, meaning one chromosome copy has the variant. In contrast, the other is normal, similar to the rest of the population.

9.4.5.1 Inheritance Patterns as a Random Recombinant Process

Figure 9.3 illustrates how inheritance patterns vary depending on whether one or both parents are homozygous or heterozygous for the gene variant. The corresponding figure panel displays the likelihood of inheriting a specific gene variant. There are four possible combinations for transmitting a gene variant from the parents. One or both parents may be carriers

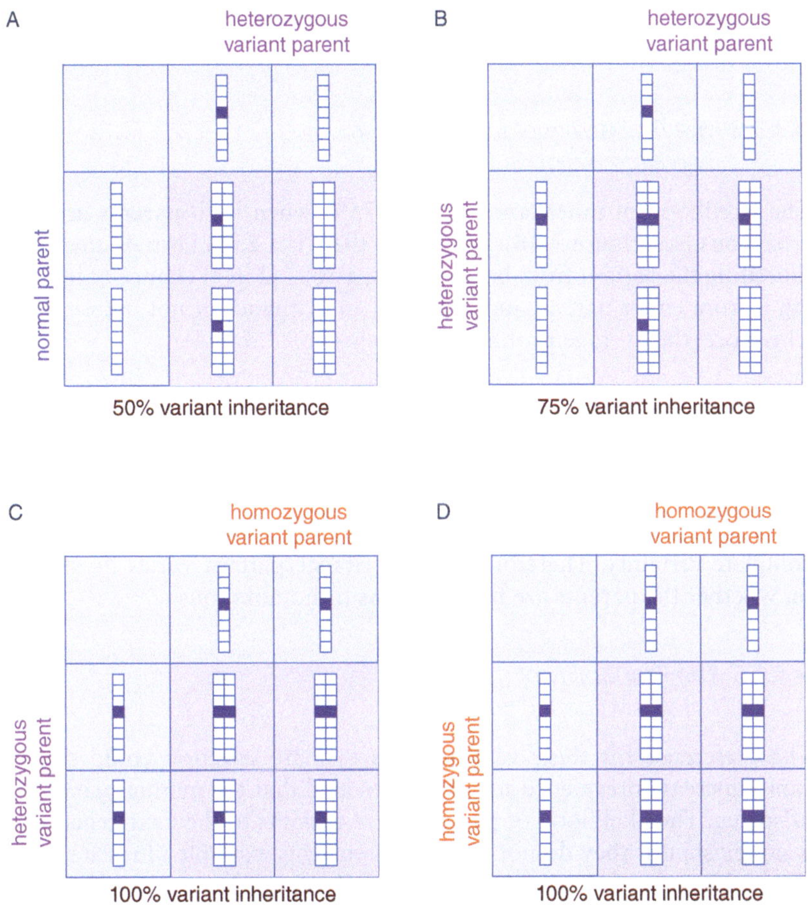

FIGURE 9.3 The table illustrates the inheritance patterns of gene variants passed on to the next generation. The probability of inheritance depends on whether one or both parents are heterozygous or homozygous. (The gene variant is indicated by a blue dot in the DNA sequence, while white dots represent the normal gene sequence.)

of the variant, either as heterozygous or homozygous carriers. Figure 9.3 shows the associated probability of inheritance.

9.4.5.2 The Probability of Inheriting from a Heterozygous Parent and a Normal Parent

If one parent is a heterozygous carrier with one copy of the gene variant and the other is normal, there is a 50% chance that the offspring will inherit the variant (see Figure 9.3A). The offspring have a 50% chance of inheriting a heterozygous gene variant.

9.4.5.3 The Probability of Inheriting from Two Heterozygous Parents

The likelihood of inheritance rises to 75% when both parents are heterozygous (see Figure 9.3B). It suggests there is a 25% (1 in 4) chance of inheriting the variant from both parents, a 50% (2 in 4) chance of inheriting it from either parent, and a 25% (1 in 4) chance of not inheriting it. Therefore, the average probability remains 75% (3 in 4).

9.4.5.4 The Probability of Inheriting from Homozygous Parents

If one or both parents are homozygous, the likelihood increases to 100% (see Figure 9.3C and D). The offspring will undoubtedly inherit it with complete certainty. Therefore, the inheritance pattern varies depending on whether the parents are heterozygous or homozygous.

9.4.5.5 The Probability of Inheritance for Parents without Children

The preference for these variant traits in mate selection could significantly increase prevalence statistics, provided that the mating pairs have offspring. The likelihood of passing these variants to the next generation is nonexistent if they do not have children. This example illustrates how mate selection and reproductive success affect population prevalence statistics beyond random chance.

9.4.6 Hardy-Weinberg Equilibrium

Genetic variation within a population will remain stable across generations if the reproductive process is entirely random. This phenomenon is referred to as the Hardy-Weinberg equilibrium. It assumes random mating and reproduction within a large population free from disruptive influences. As a result, genotype and allele frequencies will remain unchanged.

9.4.6.1 Factors Influencing Hardy-Weinberg Equilibrium

However, real-world factors such as mutations, meiotic drive, natural selection, non-random mate selection, genetic drift, gene flow, and harmful alleles can disrupt this equilibrium. These factors may influence the estimated probability, especially when a conditional probability depends on a prior probability. Conversely, when tracing ancestral history, an offspring inherits one copy of each pair of gene sequences directly from each parent with certainty. Thus, the probability of inheritance relies on the direction of tracing and the analysis performed.

9.5 NUCLEOTIDE VARIANTS AS BIOMARKERS

Forensic analysis often emphasizes potential lineage over the effects of variations in gene expression. It typically selects non-coding regions that are not expected to be transcribed for polymorphism analysis. These short sequences act as biomarkers for identification rather than providing a comparison of the entire genome. A single-nucleotide base in a DNA sequence may change due to mutations, replication errors, inheritance, or other factors. Nevertheless, there is at least a 50% chance that it will be passed on to the next generation since offspring inherit one chromosome from each parent.

9.5.1 Single-Nucleotide Polymorphism

Single-nucleotide polymorphisms (SNPs) are variations of a single-nucleotide within an identical DNA sequence (see Figure 9.4). Polymorphism refers to the substitution of one nucleotide for another within a sequence. It can arise from mutations, replication errors, or other initial causes. The substituted nucleotide is then passed down to subsequent generations, leading to variations among similar individuals in a population.

9.5.1.1 Homozygous and Heterozygous Alleles

An allele is one of two or more variations of a DNA sequence (a single base or a segment of bases) found at a specific genomic location. An individual inherits two alleles—one from each parent (see Figure 9.5). If the two alleles are identical, the individual is homozygous for that allele; if they are different, the individual is heterozygous.

9.5.2 Polymorphic Sequences as Biomarkers

The biomarker sequence is frequently chosen for its polymorphic characteristics, which exhibit nucleotide variations in gene sequences among individuals within a population. Common biomarkers include SNPs, short tandem repeats, and microhaplotypes. They are the most prevalent polymorphisms.

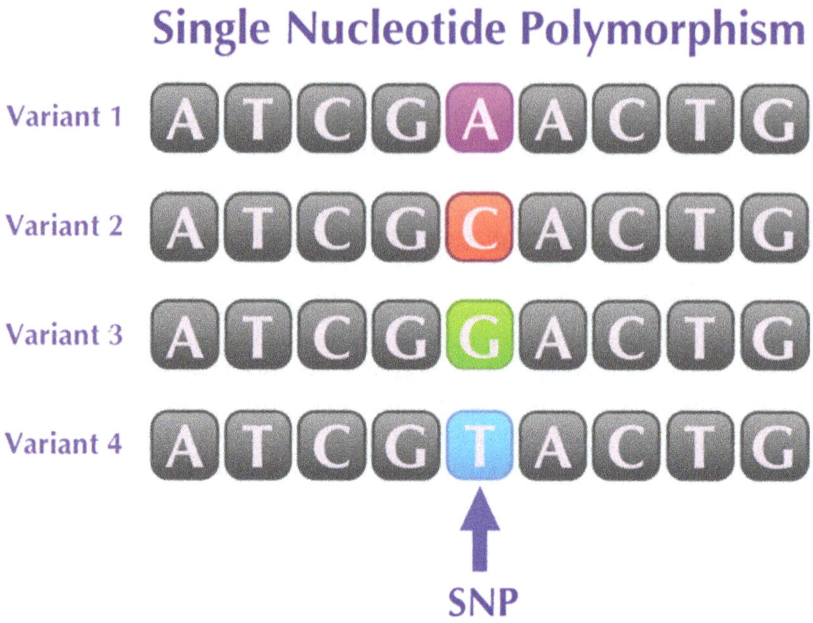

FIGURE 9.4 A single-nucleotide polymorphism (SNP) in a single-stranded DNA sequence represents a variation at a specific nucleotide position within an otherwise identical DNA sequence, differing solely at that position. There are four possible variants of an SNP, each corresponding to substituting one of the four different nucleotide bases, assuming that deletion or insertion is excluded.

9.5.2.1 Polymorphism Tracing

If the nucleotide alteration originated from a mutation in a parent, then the descendant's inheritance is unique to that parent. However, if it was passed down from previous generations, the likelihood of inheriting the polymorphism depends on its statistical incidence in the population. This probability can be estimated by comparing it to the occurrence in the target population.

9.5.3 Single-Nucleotide Polymorphism as a Biomarker

When inherited, a single-nucleotide difference can serve as a unique biomarker to trace a direct lineage from a parent instead of comparing complete genomic sequences. This similarity arises because the sequence matches the entire population, differing by only a single-nucleotide

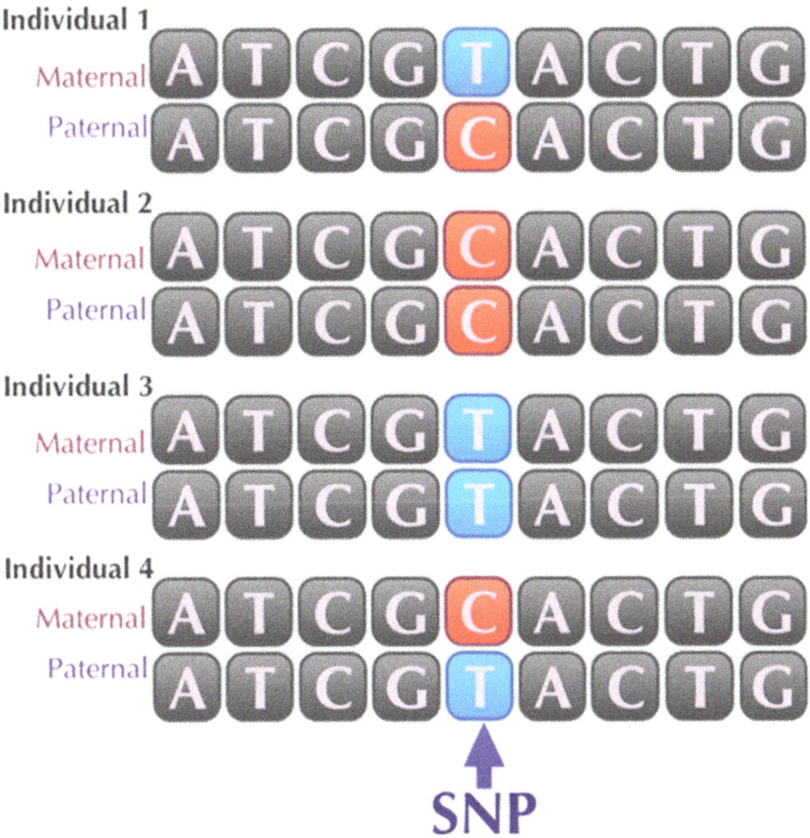

FIGURE 9.5 A single-nucleotide polymorphism (SNP) in a double-stranded DNA sequence. One strand is inherited from the mother, while the other comes from the father. This diagram illustrates a variation in a single-nucleotide within an otherwise identical DNA sequence across four individuals.

passed down from that ancestor during ancestry tracing. Even a single-nucleotide variation can distinctly identify an individual based on these DNA variants.

9.5.3.1 Comparison with a Target Population

Suppose a single base substitution differentiates an individual's DNA sequence from the general population. In that case, it may be inherited from the mother, father, or both parents since each chromosome strand originates from one parent. This variation aids in identifying descendants from a specific subpopulation and provides insights into their ancestral characteristics. Additionally, it can help identify individuals based on inherited patterns. For analysis, the likelihood of inheriting a particular polymorphism is compared to its frequency in the target population.[6]

9.5.4 Case Example: Tracing the Ancestry of Domesticated Dogs

The methodologies for tracing the ancestry of dog domestication are illustrated through a case study. While it is widely accepted that gray wolves are the ancestors of dogs, there is no consensus on when, where, and how this transition took place. This case study will demonstrate the analytical principles used to trace the ancestral origin.

9.5.4.1 Tracing the Distribution of Ancient Wolves Across Time and Space

Archaeological evidence from skeletal remains indicates that modern dogs first appeared approximately 14,000 years ago. Genetic data show their divergence from wolves occurred between 40,000 and 14,000 years ago. Identifying genetic diversity in wolves over time and across regions may clarify which populations were most closely related to the ancestors of dogs.

9.5.4.2 Tracing Ancestral Origins Using SNP Biomarkers

SNP genotypes were collected from a dataset covering the past 100,000 years, including data from contemporary wolves, ancient dogs, and various canids. An ancient dhole genome served as a control outgroup.[7] The analysis indicated that 69% of the wolves were male, based on X-chromosome DNA sequencing from 66 genomes across Europe, Siberia, and northwestern North America. This finding corresponds with a similar overrepresentation of males observed in the genomes of ancient woolly mammoths, bison, brown bears, and domestic dogs. If ancient dogs had different affinities with wolves before domestication, then gene flow from dogs could not have impacted wolves.

9.5.5 Principal Component Analysis

Principal component analysis (PCA) uncovers hidden factors in datasets by identifying essential variables. This versatile tool is applicable to various data types, revealing latent variables as linear combinations of the original ones. PCA identifies the main contributing factors by rotating the principal component axes to minimize deviation. This iterative process continues until the remaining variables contribute negligibly. The first few components capture the most variance in descending order, thereby reducing dimensionality by excluding variables that contribute minimally. The new variables simplify analysis without compromising data fidelity. Although abstract variables capture the essence of the data, they may lack direct physical interpretation unless connected to recognizable phenomena such as thought or consciousness. These variables emerge from quantifiable neural processing that PCA reveals.

9.5.5.1 Data Analysis of Unidentified Latent Factors

PCA is an analytical method for exploratory analysis that identifies latent factors influencing results. This technique provides unbiased insights into datasets. If unsure which factors affected your data, PCA can reveal them by analyzing variables through new combinations to highlight significant ones. Even without prior knowledge of the factors, this method uncovers them using a best-fit model, yielding insightful outcomes.

9.5.5.2 Case Example: Abstract Factors Revealed by PCA

PCA creates abstract factors from multiple variables, streamlining complex data into more straightforward representations. For instance, census data on income, education, rent, car value, and postal code may seem unrelated. The data exists in a five-dimensional space with five variables, making visualization challenging. PCA addresses this by generating a single factor that combines all five variables into one, positioning data points along an axis that best fits. It identifies an abstract "socioeconomic" axis that aligns with the data, effectively reducing the five variables to one and clarifying the impact of a single factor on society more effectively than the original variables did.

9.5.5.3 The Versatility of Principal Component Analysis

Unlike AI tools that often lack explanations, PCA uses statistical models to represent data across new dimensions of contributing variables. This versatile technique applies to most data types. It offers statistical transparency, reliability, and reproducibility, even for those with limited statistical knowledge. It uncovers unknown factors without prior knowledge of the contributing variables that govern the data.

9.5.5.4 Factors Identified by PCA

The major factors are the principal components that account for most of the overall variance, while the remaining factors are insignificant as they contribute little to that variance. Mapping the data into the principal components reduces dimensionality while preserving the essence of the data. When PCA extracts features using a variable, the principal component axes represent the feature vectors. When it describes the variations in the data through an equation, that variable becomes a feature, with the feature vector pointing toward the principal component axes.

9.5.6 Pearson Correlation for Identifying Population Clusters

When PCA is applied to a matrix representing shared genetic drift, it indicates that ancient wolves cluster by age rather than geography. This finding is corroborated by the Pearson correlation coefficient, which measures the linear relationship between datasets. The Pearson correlation assesses the similarity between two datasets by comparing their attributes, producing a score that ranges from −1 to +1. A high score signifies strong similarity, while a score close to zero indicates no correlation.

9.5.7 Gene Flow Directionality Analysis

The analysis of gene flow directionality indicated that Siberian ancestry spread to Europe via the Bering land bridge from Alaska and the Yukon around 10,000 years ago, but not vice versa. This allows for the deduction of ancient wolf migration patterns.

9.5.8 Maternal Lineage Analysis Using Mitochondrial DNA

Mitochondrial DNA is inherited solely from mothers, making it a valuable marker for tracing maternal lineage. Analyzing maternal mitochondrial DNA inheritance sheds light on the ancestry of wolves as they migrated from Siberia to Europe. Maternal inheritance acts as a powerful tool for defining inheritance patterns.

9.5.9 Analysis of Allele Prevalence in Natural Selection

Natural selection was confirmed by analyzing allele preferences in a dataset spanning approximately 100,000 years (around 30,000 generations). The study assessed each variant regarding allele frequency and timeframe across 72 ancient and 68 modern wolves within 24 genomic regions while accounting for genetic drift to reduce false positives. The survival of Pleistocene wolves resulted in rapid adaptations in selected alleles.

9.5.10 Dual Ancestry Model

PCA rejected a single ancestor for Near Eastern dogs. A wolf admixture model similarly confirmed varying ancestry proportions that accounted for asymmetries. The dual ancestry model indicates that Arctic dogs experienced less Western influence, while the Western components in Near Eastern and African dogs are associated with Near Eastern wolves.

9.5.10.1 Dual Origins of Ancestral Trees

The conclusion suggests two independent domestication events: one from Eastern and Western ancestors and another from a separate occurrence in the East, accompanied by Western admixture across multiple lineages.[7] Therefore, the assumption that a single origin must be identified should be cautiously approached in genealogical tree analysis.

9.5.11 Considerations in Hypothesis Testing

The above example illustrates the importance of using appropriate analytical techniques to draw valid conclusions. Misleading results can occur without the proper analytical tools or assessments to address underlying assumptions.

If the hypothesis is based on a single lineage assumption, it may overlook potential evidence of dual ancestry. Dog domestication likely happened independently among various geographical populations due to their ability to respond to commands and assist in hunting during the evolution of human hunter-gatherers.

9.5.11.1 Self-fulfilling Prophecy

Proving a hypothesis with supportive evidence is insufficient for concluding and requires analysis that challenges the hypothesis. A hypothesis is only valid if it cannot be disproven. Proving a conjecture without disproof can lead to a self-fulfilling prophecy. Proving a conjecture without disproof reflects a subconscious bias when the proof focuses solely on validating evidence without considering its potential invalidity.

9.5.11.2 Challenging the Validity of Assumptions

This caveat often occurs when conclusions depend on preconceived notions assumed to be validated before any analysis. Consequently, the analysis selectively incorporates only supporting evidence while disregarding contradictory evidence. These represent subconscious biases in any analysis that one should be aware of.

9.5.11.3 Validating Hypothesis Testing Without Refuting It

The common fallacy in hypothesis testing is presenting only evidence that supports the hypothesis while failing to disprove it. A hypothesis is considered proven only if it cannot be refuted. Confirming its validity without showing that it might be incorrect reinforces a self-fulfilling prophecy.

9.5.11.4 Including an Opposing Hypothesis in the Analysis

Proposing an alternative hypothesis is essential to avoiding premature conclusions. A robust hypothesis must be backed by strong evidence and a counterargument. An alternative hypothesis helps to dispel myths and offers a possible explanation if the evidence does not support the original hypothesis.

9.6 CONSIDERATIONS FOR LIMITED DNA QUALITY AND QUANTITY

Utilizing a short DNA segment for forensic analysis is particularly crucial when DNA samples are of low quality or limited quantity, as this ensures accurate human identification or paternity testing. Only a small amount of fetal DNA is found in maternal blood samples without performing an amniocentesis for prenatal paternity testing. In forensic analysis at crime scenes, a small amount of degraded DNA may be available. Consequently, specialized techniques will be essential to perform DNA analysis on limited sample sizes and quality.

9.6.1 Case Example: Non-Invasive Prenatal Paternity Testing

Conventional prenatal paternity testing involves invasive amniocentesis, which carries a risk of miscarriage during the collection of amniotic fluid or umbilical cord blood. An alternative method is to obtain fetal DNA samples from maternal blood, though the quantity is limited. Fortunately, techniques are available to analyze insufficient fetal DNA for conclusive paternity identification. Non-invasive prenatal paternity testing (NIPPT) collects small amounts of fetal DNA from maternal blood for analysis, thereby eliminating the risks associated with amniocentesis.

9.6.1.1 Limited Quantity of Fetal Fraction DNA in Maternal Blood Samples

Fetal fraction refers to the proportion of cell-free fetal DNA present in maternal plasma. It varies among individuals and is influenced by gestational age and weight. It is about 15% on average, ranging from less than 4% to over 30%. This parameter is crucial for the accuracy of cfDNA-based prenatal paternity tests.[8]

9.6.1.2 Maternal Cell-Free Fetal DNA Analysis

Maternal cfDNA paternity testing typically utilizes STR or SNP genotyping. The increased polymorphism of STRs benefits forensic Combined Paternity Index (CPI) and Cumulative Probability of Exclusion (CPE) analyses. However, cfDNA sizes ranging from 140 to 160 bp may exceed STR coverage, requiring additional techniques.

9.6.1.3 Improved Analytical Models for NIPPT

Models that incorporate fetal fraction and genotype probability improve the evaluation process. The Poisson-based fetal fraction model offers accurate estimates of fetal fractions without depending on known biological parents. Meanwhile, the genotype probability model enhances statistical power by merging the Poisson distribution with sequencing error rates. New CPI and CPE models provide more precise estimates based on NIPPT characteristics. Additionally, a T-test model can identify sample contamination due to its sensitivity to abnormal data.[8]

9.7 VARIABLE NUMBER TANDEM REPEATS AS BIOMARKERS

In addition to SNPs, other types of DNA variants exist as repeated patterns in succession. These variants often comprise short sequences of repeats on the chromosome that vary from one individual to another. The number of tandem repeats can differ within the general population.

9.7.1 Variable Number Tandem Repeats

Short tandem repeats (STRs) are short DNA sequences made up of tandemly repeated units of 1 to 6 base pairs (bp), leading to sequences that can differ in length by up to 100 nucleotides (nt) (see Figure 9.6). Due to the variation in the number of repetitions among individuals, they are

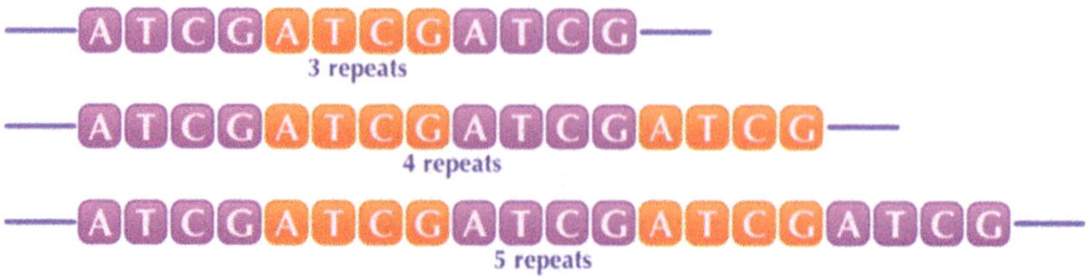

FIGURE 9.6 Short tandem repeats (STRs) are short, identical DNA sequences that repeat several times consecutively.

also known as variable number tandem repeats (VNTRs). Furthermore, they are also called microsatellites or simple sequence repeats.

9.7.2 Short Tandem Repeats as Biomarkers

The frequency of STRs is significantly higher in the general population. These tandem repeats comprise over 50% of the human genome and notably exceed the prevalence of SNPs.[9] Consequently, STRs are often employed as biomarkers to monitor inheritance patterns due to their widespread presence in the general population.[10]

9.7.3 Multiple-Locus Variable Number Tandem Repeats as Biomarkers

Multiple-locus VNTR fingerprinting (MLVF) employs VNTRs from various loci as biomarkers for identification. It requires calculating the repeat numbers for each locus. However, it does not allow for a straightforward and unambiguous determination of individual repeat counts at each locus. The primary drawback of this method is the absence of direct result comparisons among laboratories. Conventional electrophoresis on low-resolution agarose gels presents amplicons only as banding patterns and does not provide accurate repeat counts or correlations to PCR targets.[3]

9.8 EPIDEMIOLOGICAL STUDIES CONNECTING STRs TO DISEASES

Epidemiological studies often clarify the genetic link to human diseases through disease-associated STRs. Examples include the genome-wide search for common STRs related to the genetic risk of Parkinson's disease[11] and the variants of the VNTR of the DRD4 gene associated with novelty-seeking personality traits and ADHD (attention-deficit hyperactivity disorder). The number of tandem repeats can trace the ancestral origin of these variants linked to ADHD and determine whether their prevalence results from chance or selection.

9.8.1 Case Example: Linking DRD4-7R VNTRS to ADHD

When the number of repeats is exactly seven, the 7-repeat phenotypic expression of the DRD4-7R allele is strongly associated with personality traits like novelty-seeking and ADHD among European Caucasians, South Americans, and individuals of Middle Eastern descent.[12] In the general population, the number of tandem repeats for the DRD4 gene, which codes for dopamine D_4 receptors, ranges from 2 to 11.

9.8.1.1 The Number of DRD4 Repeats in the General Population

The number of repeats in the general population ranges from 2R to 11R for the 48-bp VNTR variants of the DRD4 gene, situated near the telomere of chromosome 11p in exon 3, which encodes the dopamine D_4 receptor. The variants 2R, 4R, and 7R comprise over 90% of allelic diversity, with 4R being the most prevalent, whereas the frequencies of 2R and 7R vary significantly depending on geographic location.[13, 14]

9.8.2 The Use of Homozygous Variants to Trace Ancestral Histories

The frequency of occurrence can trace specific ancestral events that lead to the inheritance of identical gene variants from both parents. For the DRD4 gene, the most common homozygous variants are 2R/2R, 4R/4R, and 7R/7R, which help determine whether these specific variants arise by chance or through non-random selection. This polymorphism also assists in determining allele ages, indicating when the genetic lineage began to diverge.

9.8.2.1 Tracing the Divergence of the Ancestral 2R, 4R, and 7R Variants

In other words, one can trace the events that led to the divergence of the 2R, 4R, and 7R variants within specific subpopulations in certain geographic regions. This outcome also highlights the migration patterns in human history that have shaped the current demographic distribution. This finding is a clear example of the genealogical methodologies and analyses required to trace the ancestral events that caused the divergence of the genetic traits of the DRD4 gene variants.

9.8.2.2 Linkage Disequilibrium Analysis

Linkage disequilibrium (LD) refers to the non-random association of alleles at different loci, serving as a sensitive indicator of the population genetic forces shaping a genome.[15] The LD between alleles and traits facilitates fine-scale gene mapping through genome-wide association studies, aiding in identifying SNPs linked to complex diseases.[15]

9.8.3 Natural Selection or Random Occurrences

No single statistic effectively quantifies LD; however, local and genome-wide LD patterns provide insights into natural selection and historical population dynamics.[16] Recombination patterns can reveal complex interactions among selection, mutation, and genetic drift, all of which influence LD levels. High local LD indicates a recently favored allele under strong selection, while low LD suggests random selection without preference in mate choice. LD patterns can aid in identifying selected loci and estimating allele ages.[15]

9.8.3.1 Evaluating the Selection Preferences of 2R/2R and 4R/4R

Drawing on ancestry from Africa, Europe, Asia, North and South America, and the Pacific Islands, the data for 2R/2R and 4R/4R homozygotes did not achieve statistical significance. It showed minimal LD. Most other prevalent DRD4 VNTR variants do not demonstrate a selection preference, suggesting that parental mate selection is random for these 2R/2R and 4R/4R alleles.

9.8.3.2 Evaluating the Selection Preferences of 7R/7R

In contrast, the 7R/7R homozygote data revealed statistical significance through Tajima's D test,[17] highlighting a strong LD at most polymorphic sites. The evidence for robust LD surrounding the 7R allele is compelling, as all 7R/7R individuals (including those from Africa) exhibit a strong selection preference from both parents.

9.8.3.3 Evaluating the Recombination Pattern of 7R/7R Selection

The recombination pattern of the homogeneous 7R/7R allele suggests that selection plays a significant role in inheriting identical gene variants from both parents. The 7R variant likely originated from a rare mutation before becoming widespread due to positive selection, indicating that this selection was not a chance occurrence.

9.8.3.4 Selection of Mating Pairs for 7R Variant Traits

Evidence suggests that strong selection has raised the allele frequency to levels surpassing those expected from random genetic drift. The selection of 7R traits by mating pairs may result in higher birth and survival rates than other pairs. Beyond random selection, the greater prevalence of the 7R variant in human populations likely arises from positive selection and the survival advantages associated with this variant's gene expression.

9.8.4 The Non-Random Selection Reflected by Homozygous 7R/7r Inherited from Both Parents

The non-random selection suggests evolutionary advantages that enhance survival rates in phenotypic traits such as novelty-seeking and ADHD associated with the 7R variant. This variant may confer benefits that have persisted despite the detrimental effects of a defective gene variant on regulating impulsive behaviors due to the D_4 receptors' insensitivity to dopamine signals.

9.8.4.1 The Survival Advantages of Non-Random Selection for DRD4-7R Traits

Dopamine's role in suppressing choices during decision-making is crucial for managing impulses against distracting stimuli. This regulation of decision-making distinguishes humans in their intellectual evolution. However, the resulting dysregulation, which leads to impulsive and novelty-seeking behaviors, paradoxically increases survival rates compared to the general population. Novelty-seeking may have encouraged human ancestors to migrate, explore other continents, and reduce resource competition during their migrations, thereby enhancing the group's survival rate.

9.8.5 The Common Disorder-Common Variant Hypothesis

The common disorder-common variant hypothesis (CDCVH) posits that if a heritable disease is prevalent, with a prevalence exceeding 1% to 5% in the population, its genetic contributors will also be widespread.[18] Genetic markers are identified in variants of coding and regulatory regions. This hypothesis applies to specific variants that increase susceptibility to complex polygenic diseases, with each variant contributing a small additive effect that combines into multiplicative effects on disease phenotypes.[18]

9.8.5.1 The Paradoxical Benefits of DRD4-7R in ADHD as a Disorder

The link between the 7R allele and ADHD suggests that both environmental and genetic factors may influence this prevalent disorder. The statistical prevalence of ADHD in the general population ranges from 3% to 5%. Traits associated with DRD4-7R could predispose individuals with ADHD to behaviors that may be harmful in certain environments but beneficial in others. While the CDCVH indicates deleterious effects related to a genetic disorder, the 7R traits might promote an evolutionarily advantageous strategy that results in positive outcomes rather than the negative behavioral consequences typically associated with the common disorder.

9.8.5.2 Single-Step Mutation Random Events for Most DRD4 Variants

Polymorphic variation is observed across 67 haplotype variants. Most haplotype variants, such as 2R-6R, involve a single variant resulting from a one-step mutation. These likely represent random occurrences due to mutations or recombination errors.

9.8.6 Multi-Step Mutation Event Pattern for DRD4-7R Variants

The DRD4-7R variant shows a pattern associated with nearby polymorphisms. The connection between 7R/7R homozygotes and four adjacent DRD4 polymorphisms suggests that they have likely undergone at least six mutations or recombination events. This sets it apart from other common variants that contain a single mutation.

9.8.6.1 Evidence of a More Recent Origin for Multi-Step Mutation in DRD4-7R

Although rare, this pattern likely spread through non-random mate selection for specific traits, resulting in a higher prevalence than random selection for other alleles. This prevalence prompted an examination of the LD between the 4R and 7R alleles. The rare multi-step mutation in 7R/7R homozygotes probably occurred more recently than in other homozygotes, such as 4R/4R, which underwent a single one-step mutation. The recombination pattern at these polymorphic sites aligns with the selection pattern at DRD4-7R.

9.8.7 Allele Age Calculations

Estimating the age of an allele offers insight into when the variant first emerged. This equation summarizes the standard methods for calculating allele age, as represented in generations:

$$t = \frac{1}{\ln(1-c)} \ln \frac{x(t)-y}{1-y} \tag{9.1}$$

where c is the recombination rate, $x(t)$ is the frequency in generation t, and y is the frequency on ancestral chromosomes. The age of the allele is calculated using the formula above when the recombination rate c is available, and $x(t)$ and y are derived from the genetic survey data.[19]

9.8.7.1 Dating the Ages of Most Other DRD4 VNTR Variants

Calculating the allele age based on the high global prevalence of DRD4-2R, 4R, and 7R suggests that these alleles have an ancient origin, estimated to range from approximately 300,000 to 500,000 years.[14] However, the branching of the common ancestor for these variants differs in allele age.

9.8.7.2 Determining the Age of the Divergence of DRD4-7R from Its Common Ancestor

Strong LD exists between the 7R allele and nearby DRD4 polymorphisms, suggesting that it is 5 to 10 times younger than the common 4R allele. Based on the variability of 18 observed intra-allelic heterozygosity sites across the locus, the age of the 7R allele is estimated to be approximately ten times younger. An analysis of high-heterozygosity sites indicates that the most recent common ancestor of the 7R/7R allele emerged between 40,000 and 50,000 years ago.[13]

9.8.7.3 Likely Mate Selection Preferences of DRD4-7R Variants

The divergence from their common ancestors likely predates human migration to other continents during the Upper Paleolithic period.[14] Traits such as novelty-seeking may offer survival advantages by encouraging adventurous migration to areas with less competition for resources and mate selection. Their mate selection preferences likely reinforce themselves when both parents carry the variant traits, ensuring these traits are passed down to their descendants. With a subpopulation of either homozygous or heterozygous carriers isolated from the general population, the frequency of allele prevalence could quickly exceed chance levels due to these self-reinforcing conditions.

9.8.8 The DRD4 Gene Encodes the Dopamine D4 Receptor

The functional significance of the number of repeated sequences lies in the changes to the dopamine D_4 receptors (DRD4) protein within a region that couples with G proteins and mediates intercellular cAMP levels.[20] Consequently, the expression of the 7-repeat gene leads to the insensitivity of dopamine D_4 receptors to inhibitory signals during neural processing. This 7-repeat allele of DRD4-7R is associated with the expression of novelty-seeking personality traits and the prevalence of ADHD.

9.8.8.1 D_4 receptors in Executive Function Control for Task Completion

D_4 receptors are predominantly localized in prefrontal cortex (PFC) neurons, where they play a crucial role in regulating behavior and evaluating the importance of choices during decision-making. These receptors in postsynaptic neurons respond to inhibitory dopamine signals released from presynaptic neurons by binding to them, thereby suppressing alternative options in the decision-making process. This mechanism is part of the executive functions that help control impulsive behavior, which can lead to poor choices. The executive function process involves the interaction of cognitive, behavioral, and emotional regulation for making appropriate decisions, much like a CEO's role within the brain. These receptors manage selective attention by focusing on problem-solving while maintaining working memory to keep track of essential intermediate steps for effective planning toward achieving a desired goal.

9.8.8.2 The Role of Inhibition in Decision-Making Processes

A decision involves choosing one option from several alternatives. Completing a complex task requires decision-making to identify the most crucial options and working memory to retain temporary information vital for task completion. Neurons in the PFC prioritize these options based on their importance, facilitating the selection of the top choice on the list. However, if some options are equally important, a decision must be made to resolve the tie; otherwise, one may become trapped in indecision.

9.8.8.3 Conflicts in Decision-Making

A conflict arises when options are mutually exclusive; choosing one negates the other. For instance, a conflict occurs when you want to have your cake and eat it too. Resolving a conflict requires setting aside competing options to make a desirable choice. It involves suppressing alternatives to enable the selection of the preferred option. If the other options are not properly suppressed, it will result in an impulsive decision that is incompatible with the desired solution.

9.8.8.4 The Conflict Resolution Process

Dopamine functions as the inhibitory signal that prevents the selection of undesirable options during decision-making. Presynaptic neurons in the PFC release dopamine to suppress the responses of postsynaptic neurons, enabling them to disregard irrelevant options. When D_4 receptors become insensitive due to defects in the DRD4-7R variant, they fail to respond to the inhibitory signal, leading to poor decisions and behaviors such as distraction, impulsivity, and novelty-seeking—key symptoms of ADHD.

9.8.8.5 The Pathology of Inheriting the DRD4 7-Repeat Sequences

Indecision arises when neurons fail to resolve conflicts between choices, causing them to become stuck in indecision. Impulsivity occurs when they select an inappropriate response that seems significant as a choice despite being otherwise. Distraction happens when they focus on trivial stimuli that compete for attention. This scenario resembles driving a car with malfunctioning brakes, which can lead to crashes due to failed stopping attempts. This example illustrates the consequences of variations in the number of repeated DNA sequences that affect the sensitivity of a receptor binding site, thereby influencing behavior and neural processing. However, the prevalence statistics persist, suggesting that the disorder may confer beneficial effects on survival, in contrast to the assumed detrimental impacts of a disorder associated with novelty-seeking and curiosity behaviors, which are crucial for innovation, discovery, and migration away from resource competition — factors that likely provide a competitive advantage in survival.

9.8.8.6 The Unforeseen Impact of a Gene Variant on Global Migration and Technological Innovation

This case example illustrates that tracing ancestral lineage has revealed unexpected results from inheriting a gene variant that may confer beneficial effects despite the behavioral deficits of impulsivity and distraction. The timing of the ancestral gene variant divergence coincided with the onset of global migration, which preceded the subsequent discovery of the New World and the innovations associated with the technological advancements that began during the Upper Paleolithic.

9.9 MICROHAPLOTYPES CONSIST OF TWO OR MORE SNPs

Microhaplotypes (MHs), often called microhaps, are chromosomal segments that typically consist of two or more closely located SNPs at the molecular level (less than 300 bp).[21] When two or more SNPs are positioned near each other, their alleles can be inherited together from parent to child. The variants often cluster together.

9.9.1 The Use of Microhaplotypes as Biomarkers

Because of the short lengths of microhaplotypes, these sequences are often used as genetic markers. They are called "micro" because they cover only a few dozen base pairs instead of thousands. Microhaplotypes (MHs) have proven to be highly informative in demonstrating the uniqueness of DNA profiles and determining an individual's biogeographic ancestry. Biological relationship tests like paternity tests frequently use MHs as biomarkers.

9.9.1.1 The Use of SNPs, STRs, and MHs as Biomarkers

Although the polymorphism analyses for each type of biomarker are similar regarding their occurrence frequency within the target population, notable differences arise in the sequences of variations.[6] The SNP variants are limited to four types, as only four nucleotide bases can substitute:

A, C, G, and T. Thus, there are four SNP variants. In contrast, the STR variants depend on the number of repeats, with probabilities of occurrence based on this count within the sequence. As the name suggests, VNTR indicates a variable number of tandem repeats, and the repeated sequence varies among individuals in the population. Identifying the number of repeats in a sample may not be precise when using amplification techniques. It is essential to recognize that false positives may arise due to sample contamination, which usually results from carryover in prior DNA analyses.

9.9.1.2 The Use of Non-Coding Regions in Forensic Analysis

Forensic analysis focuses on connecting individuals rather than solely emphasizing gene expression or the effects of protein synthesis from coding regions. Excluding coding regions from DNA sequences during forensic analysis can simplify the complexities of phenotypic expression, which may impact the results. Therefore, polymorphisms in non-coding areas are typically preferred for DNA analysis. Most DNA sequence databases used in forensic analyses primarily consist of non-coding regions.

9.10 THE FORENSIC COMBINED DNA INDEX SYSTEM DATABASES

The Combined DNA Index System (CoDIS) is a national DNA index system consisting of a collection of databases developed and maintained by the United States Federal Bureau of Investigation (FBI) to collect data on various STRs. This system distinguishes individuals based on the lengths of these alleles. The database uses a set of 20 loci tested together to improve the reliability of the analysis.[22] A locus refers to a specific location on a chromosome for a particular gene or genetic marker. The twentieth locus in CoDIS, known as AMEL, is used to identify the sex of individuals based on the sex chromosomes.[6]

9.10.1 The Use of 20 loci as an Index for Identification

Using 20 loci significantly reduces the risk of confusing two distinct individuals, as individuals often share alleles at specific loci, especially when related. It is analogous to using physical characteristics for identification: the more traits used to describe a person, the less likely they are to be mistaken for someone else.

9.10.1.1 Forensic Index System Databases

The Forensic Index includes profiles derived from biological evidence at crime scenes, such as blood or semen samples, which help identify remains when linked to a known source.[23] DNA profiles in each index are compared to find direct matches or potential relationships based on shared genetic data. Searches for familial relationships are conducted solely for missing persons and unidentified remains.

9.10.1.2 The Unidentified Human Remains Index

The Unidentified Human Remains index features profiles of human remains. In contrast, the Relatives of Missing Persons (RMP) index includes voluntarily collected profiles from biological relatives, typically obtained from buccal swabs but sometimes sourced from blood or other samples. These profiles are indexed in this database.

9.10.1.3 The Pedigree Tree Index

The Pedigree Tree index organizes relatives' specimens into family groups for efficient searches and is compared only to the Unidentified Human Remains index in CoDIS. Samples from missing persons are stored in the missing persons index and may include personal items such as toothbrushes or hairbrushes.

9.10.1.4 The Convicted Offender Index

The Convicted Offender Index contains profiles of individuals convicted of qualifying offenses. According to state laws, blood or buccal swabs are collected from arrestees. Some states also collect additional samples.

9.10.2 The Use of 13 Core STR Loci

A reduced number of loci is sufficient for identifying an individual without requiring all 20 loci. The FBI has identified 13 core STR loci as adequate for individual identification in the U.S. Conversely, Interpol (the International Criminal Police Organization) has determined that 10 core STR loci are sufficient for identification in the U.K. and Europe. Interpol is an intergovernmental organization composed of 196 member countries that facilitates police collaboration across these nations. In the U.S., the 13-STR profile is a standard identification method used to determine human remains, establish paternity, or link suspects to crime scenes.

9.10.2.1 Using 13 Core STR Loci for Identification

The FBI determined the frequency of each allele for the 13 core STRs across various ethnic groups by analyzing DNA from hundreds of unrelated individuals. Assuming all 13 STRs exhibit independent permutations, statistical calculations indicate that the probability of two unrelated Caucasians sharing identical STR profiles, or "DNA fingerprints," is approximately 1 in 575 trillion.[24]

9.10.2.2 The Matching Fallacy

However, this probability pertains to pairs of individuals globally. With 100 million Caucasians, there are 5,000 trillion pairs, suggesting that approximately eight or nine pairs would match at the 13 STR loci. This matching does not indicate which profiles two individuals share, and the likelihood of matching a crime-related profile remains very low.[25]

9.10.2.3 A STR Match

A laboratory analyzes the allele profiles of 13 core STRs from both samples to connect evidence from a crime scene to a suspect. If the STR alleles do not match, the individual is excluded as a potential source of the evidence. If they match all 13 STRs, a statistical calculation estimates how frequently this genotype occurs in the population, considering the prevalence of each STR allele in the individual's ethnic group. A Hardy-Weinberg calculation determines the frequency of the observed genotype for each STR, and multiplying the frequencies of the individual STR genotypes provides the overall profile frequency.

9.10.2.4 Analysis of Conditional Probabilities for STR Matching

If Suspect A is excluded as the source of the crime scene sample, and Suspect B matches all 13 STRs, the likelihood that a random member of Suspect B's ethnic group shares this genotype is extremely low. It implies the probability of observing this DNA profile if the evidence did not originate from the suspect. Misinterpreting this could confuse the probability of the suspect belonging to the ethnic group with the likelihood that the suspect is the source. Calculating the transitional probability requires Bayes' theorem and prior conditional probabilities concerning the suspect's involvement. Moreover, the probability increases significantly if a relative of the suspect is the source, particularly a sibling.

9.10.2.5 The Likelihood of a False Positive

The significance of chance phenomena cannot be overstated; even with an extremely low likelihood of a false positive, it can still occur, much like winning the mega lottery. Although the odds are one in a billion, winning is possible if you are lucky; if it never happens, no one could win the lottery.

9.10.2.6 The Confounding Variables

DNA collected from crime scene evidence is often limited in quantity, poorly preserved, or degraded, leading to partial profiles. Analyzing fewer than 13 STR loci increases the likelihood of random matches. For example, examining fewer than five initial STRs significantly raises the chances of encountering that genotype. Additional evidence regarding Suspect B should be gathered to exclude potential candidate involvement. Moreover, common STR alleles within an individual's ethnic group can raise genotype frequencies, even when all core loci are analyzed.

9.10.2.7 The Sources of Uncertainty

Crime scene samples may also contain DNA from multiple sources, complicating the analysis. Traditional forensic analysis typically relies on probabilities, which means that even a confirmed match cannot establish guilt. Additional evidence is necessary to establish the connection beyond a single source.

9.11 THE ARTIFICIAL INTELLIGENCE APPROACH

Artificial intelligence serves as an alternative method for solving complex problems through expert systems or ML capabilities. Recently, AI has gained recognition for employing ML to tackle intricate issues via examples rather than explicit instructional methods. The system discovers solutions by learning from the examples in the training dataset.

9.11.1 Machine Learning

AI provides alternatives to traditional methods for calculating conditional probabilities and establishing relationships. It effectively addresses specific inquiries in natural language by utilizing LLMs to generate suitable genealogical responses to user questions. Additionally, it categorizes historical records of both written and spoken language and census records and stores genealogical data in databases. Furthermore, ML speeds up genome analyzers, producing millions of short sequencing reads.[26] However, it is essential to recognize AI's strengths and weaknesses when evaluating the validity of conclusions.

9.11.1.1 Automation with Machine Intelligence

Today's AI is more accurately described as machine intelligence (MI) because it forms internal representations through learning instead of relying solely on preprogrammed responses. Rather than adhering to strict algorithms, these models transform internal representations from input to output using neural networks that extract information from training datasets.

9.11.1.2 Collective Neural Processing in Computing

The system generates responses using a vast network of neurons, unlike the single CPU logic found in traditional computers. AI output improves through ML, beginning with random outputs refined through extensive training on billions of examples to establish essential input-output relationships. Forming meaningful connections among data points in a dataset relies on a highly interconnected network of neurons that captures relationships by creating an internal representation of these links.

9.11.1.3 How an AI System Is Trained

An AI neural network generates outputs through interconnected tokens and connection weights. It responds to queries based on these tokens and weight matrices derived from extensive training datasets, effectively mapping inputs to outputs. Learning from billions of examples creates an internal representation of input-output relationships. The network applies learning rules to modify interconnectivity and uses criteria to identify relevant connections while discarding irrelevant ones.

9.11.1.4 The Mystery and Magic of AI Uncovered

Requesting an AI to generate images from text involves mapping inputs to outputs through training to minimize errors. This approach aids the system in adjusting its input-output connections. Pairing descriptions with images forms internal representations that enhance learning by reducing errors. Trial and error produce random outputs, which are compared to target images. A reduction in errors indicates progress, prompting the system to modify its connections accordingly. Repeating this process with a billion examples refines the image based on the description and reinforces relevant connections. The outcome is an image created from the text description.

9.11.1.5 AI Responses to User Queries

A genealogical query can yield plausible results when trained on questions with known answers, utilizing billions of examples to develop appropriate responses. Training an AI system generally requires billions of examples for the model's representation to align with expected input-output relationships. One unique feature of modern AI systems is their ability to generalize and summarize large sets of data or information.

9.11.2 How AI Produces Responses to Unfamiliar Questions

It can generate appropriate responses even if those inquiries are not included in the training dataset. It generalizes answers using internal interconnectivity to map input queries to output response relationships. This means it can provide missing information, even when the database is incomplete or the inquiries are not part of the original training examples.

9.11.2.1 Consequences of Insufficient Training

Adequate training on a large dataset can produce reliable responses. However, these responses may become inaccurate if the internal model fails to converge properly to a stable representation due to insufficient training data. An LLM model typically requires a minimum dataset size of 200 billion examples to achieve convergence toward a solution.

If the training dataset is insufficient, AI models may struggle to converge and become trapped in local minima during error minimization. Techniques are available to help escape these local minima and find better global solutions, although errors may occur while overcoming the hump. Without these methods, AI tools might inadvertently settle for suboptimal solutions.

9.11.3 Requirements for DNA Dataset in AI Training

AI for DNA analysis relies on short polymorphisms or genome-wide sequences stored in databases. Training an AI system using human genome data composed of 3 billion nucleotides is possible. Training on short polymorphic DNA sequences requires a substantial database to provide sufficient example datasets. GenBank is an open-access, annotated repository of nucleotide sequences and their corresponding protein translations. It includes 34 trillion base pairs from over 4.7 billion sequences and covers over 580,000 protein species. It contains 1.42 million SNPs, averaging one every 1.9 kb. Approximately 60,000 SNPs are present in exons (both coding and untranslated regions), with 85% located within 5 kb of the nearest SNP.[9]

9.11.3.1 Exploring Relationships in DNA Datasets

AI genealogy and forensic analyses depend on accessible datasets. When searching for candidate genes associated with diseases or drug responses, reasonable suggestions are accepted even without thorough justifications. Researchers investigate potential interactions of gene expressions using extensive SNP datasets or genomes, refining their search for additional human analysis. Such tasks are labor-intensive and time-consuming. AI can accelerate the discovery of possible gene combinations, as the matching criteria are less stringent than those used for tracing an individual.

9.11.3.2 Collective Neural Processing in Computing

AI generates human-like responses by analyzing vast amounts of data; however, it falls short in logical reasoning and genuine creativity. Although this process can be automated, it lacks true intelligence, reasoning skills, and a fundamental understanding of the world. It produces content based on training datasets, which can lead to absurd outputs resulting from misrepresentations of physical phenomena.

9.12 THE ARTIFICIAL INTELLIGENCE COMPUTING APPROACHES

AI has evolved into two approaches to automated computing, each based on distinct methodologies for problem-solving. The term "AI" originated as a theoretical concept for information processing in the 1950s, a function that was once exclusive to humans before the emergence of automated computing machines. A historical case study will demonstrate how AI has advanced from the 1950s to modern-day AI.

9.12.1 Case History: The Turing Machine

Alan Turing was a brilliant mathematician recognized as a pioneer in laying the foundation for modern computing. In 1936, while at the National Physical Laboratory in the U.K., he published a design for the ACE (Automatic Computing Engine),[27, 28] an early version of modern computers. He proposed using mechanized computation to automate a series of programmable instructions known as the "Universal Turing Machine."[29]

9.12.1.1 The Predecessors of Modern Computing Machines

In perspective, the slide rule was a mechanical device used to calculate parabolic trajectories while accounting for recoil on naval ships by solving the differential equations of its time. Mechanical devices were employed to encode and decode wartime messages during WWII. The Colossus computers, modeled after the ACE, operated from 1943 to 1945 and were used for the cryptanalysis of the Lorenz cipher. As the first programmable computers, they utilized vacuum tubes to process Boolean logic operations. The renowned EDVAC, designed by John von Neumann in 1945, was based on Turing's theoretical work.[30]

9.12.1.2 The Turing Test for Artificial Intelligence

AI dates back to 1950 when Alan Turing introduced the "imitation game" as a test for MI in his paper "Computing Machinery and Intelligence."[31] The test, known as the "Turing Test," poses the question, "Can a machine think?" If a machine's responses are indistinguishable from those of a human, it is considered intelligent. However, this test evaluates performance rather than cognitive ability. For example, a person with schizophrenia might be mistaken for a machine due to irrational responses, indicating that mimicking human interaction does not equate to true intelligence. Similarly, a zombie can imitate interactions but lacks conscious intelligence.

9.12.1.3 Machine Intelligence as an Automated System

The Turing Test for thinking machines has sparked philosophical discussions that extend beyond intelligence to encompass consciousness. Ultimately, the criteria for imitation are insufficient to differentiate consciously thinking machines from those that are unintelligent. Intelligence requires thinking, while imitation does not. The Turing Test fails to address this crucial aspect, leading to ongoing debates about intelligence that are both unnecessary and unproductive—debates that persist even today.

9.12.2 Machine Learning Without Preprogrammed Algorithms

MI refers to automated systems that use ML to process information without predefined algorithms, achieving performance levels similar to human interaction. ML emphasizes computational processes rather than outcomes. AI encompasses abilities that can surpass human capabilities.

9.12.2.1 Decoding Messages: A Prelude to AI

Turing worked at Bletchley Park in London during WWII, focusing on cracking the Enigma machine code used by German forces.[32] Initially, Polish mathematicians decrypted Enigma messages; however, the Germans changed the cipher daily. He and others at Bletchley conducted cryptanalysis using the Bombe machine and the Banburismus technique to decode messages. They successfully deciphered encrypted naval communications from German U-boats and the Air Force in 1941. In July 1942, he developed the Turingery technique to decrypt the top-secret German communications utilized in the Lorenz cipher machine.

9.12.2.2 The Demand for Artificial Intelligence in Encryption

He traveled to the U.S. in December 1942 to share his expertise in Enigma encryption and advise U.S. military intelligence. This decryption was crucial in directing Allied convoys away from the U-boat wolf packs, significantly changing the Battle of the Atlantic in favor of the Allies during WWII.[32]

9.12.2.3 Post-War Recognition of Decryption Technology

During his visit, Turing learned about the U.S. speech encoding system and developed his speech-scrambling device, "Delilah." In 1945, he was awarded the Most Excellent Order of the British Empire (OBE) for his contributions during the war. Established by King George V in 1917, the OBE honors achievements in the arts, sciences, charitable work, and public service in non-combat roles.[33]

9.12.2.4 The Official Apology

In 1952, Alan Turing was prosecuted for homosexuality, which was illegal in Britain at the time. To avoid imprisonment, he accepted chemical castration and had his security clearance revoked. In 1954, he died from cyanide poisoning at the age of 41, a death ruled as suicide. Turing's role in breaking the Enigma code remained classified until the 1970s, with the complete story emerging in the 1990s. His conviction was overturned posthumously in 2013, 60 years later, and he received a Royal Prerogative of Mercy pardon from the Queen.

9.12.2.5 Turing Award for Excellence in Computing

Turing's contributions to computer science are recognized through the annual Turing Award, which has been the highest honor in the field since 1966 and is regarded as equivalent to a Nobel Prize. Notably, the Nobel Foundation did not establish a category for computer science awards in 1895.[34]

9.13 TWO DISTINCT AI APPROACHES FOR SOLVING COMPLEX PROBLEMS

As AI has developed through advanced computing algorithms capable of solving complex problems, two fundamentally different approaches have emerged. The traditional approach employs preprogrammed algorithms to outline specific methodologies for addressing expert system problems. Conversely, the alternative approach utilizes ML to uncover solutions to complex issues without specifying the precise methods needed to solve them.

9.13.1 Traditional Algorithm Method

This approach utilizes symbolic processing and algorithms to automate tasks in expert systems. Traditional DNA analysis employs algorithmic methods within expert systems to calculate conditional probabilities through database comparisons. This illustrates the classical approach to problem-solving by distinctly outlining the methodologies for computer automation.

9.13.1.1 Special-Purpose Machines

This methodology addresses problems by following the algorithm's explicit instructions. The intelligence resides with the programmer, not the machine itself. Expert systems are designed to address well-defined issues in specific scenarios within a targeted domain. They are not meant to serve as universal problem solvers for other issues and do not generalize solutions to different problems.

9.13.1.2 Limitations Imposed by Its Preprogrammed Algorithms

Using programmed logic to tackle complex problems may seem straightforward and effective, but these algorithms limit their problem-solving capabilities. Unforeseen circumstances that surpass the programmed logic remain unaddressed unless the algorithms are modified. They cannot incorporate new algorithms or analyses without reprogramming.

9.13.1.3 Unresolvable Missing or Incomplete Information

Another limitation is its inability to address information gaps when the database is incomplete, missing, or contains biased data. The programmer serves as the primary source of intelligence rather than the machine. The machine executes automated instructions without the intellectual capacity to generate new algorithms or independently rectify errors.

9.13.1.4 The Unresolved "Hard Problems" in Natural Language and Image Processing

For decades, computers have struggled to recognize speech accents in automated phone answering systems and transcribe voicemails. Facial recognition, handwriting recognition, and natural language processing exemplify the limitations of traditional AI. These shortcomings persist because traditional AI relies on preset algorithms that cannot comprehend anything beyond their programming. Furthermore, chatbots, which do not need to process accents, cannot interpret context or extract relevant information from users' inquiries to provide appropriate responses.

9.13.2 Neural Network Machine Learning Techniques

The neural network approach uses implicit methods to derive solutions from examples without depending on explicit algorithms. The Perceptron, an early problem-solving machine, classifies images by learning from training data. Modern AI systems utilize similar learning rules and neural network architectures from the 1980s, scaled and trained on datasets of 100 billion, enabling them to address real-world challenges rather than merely solving toy problems.

9.13.2.1 Problem-Solving Using Neural Networks

Recently, ML techniques that utilize neural networks have greatly enhanced natural language processing and image recognition. These methods enable systems to accurately route calls to the appropriate department for human responses and identify faces, handwriting, and speech accents through extensive training on datasets.

9.13.2.2 General Purpose Problem Solvers

ML acquires and generalizes solutions to various problems, establishing itself as a universal problem solver. It can also address gaps created by missing data not included in the training set. This system is trainable and adapts to the training datasets. With additional training, it can refine its responses; thus, it is an adaptive system that gradually aligns with the user's environment.

9.13.3 Principles of Neural Integration: From Multiple Inputs

Neural networks adjust their weights based on training feedback and compute outputs across multiple neurons. They employ parallel processing to generate solutions from a group of neurons, relying on statistical properties instead of the accuracy of each neuron. This contrasts with the serial processing found in traditional CPUs, which utilize algorithmic problem-solving.

9.13.3.1 Learning Rules for Adjusting Connection Weights

Learning rules decrease errors or employ the auto-associative Hebbian learning rule, reinforcing active connections to improve network performance. The learning rule used during training can be either a supervised approach (learning with a teacher) or an unsupervised approach (learning without a teacher). Incremental weight adjustments enable the system to find solutions that minimize errors; therefore, these solutions arise automatically through error minimization.

9.13.3.2 Limitations of Single-Layer Neural Networks

The original Perceptron network architecture was overly simplistic, relying on a single-layer design to solve problems. This limitation caused researchers in the traditional algorithmic AI group to underestimate its problem-solving abilities, claiming that it could not tackle real-world challenges. However, this limitation was soon recognized as an inherent flaw of a single-layer neural network. In contrast, a multi-layer network architecture can tackle problems that a single-layer network cannot solve.

9.13.3.3 Overcoming Limitations Using Multi-Layer Neural Networks

Subsequently, it was found that adopting a multi-layer network architecture was crucial for addressing these complex issues. Multi-layer network models effectively overcame the limitations of single-layer neural networks. They tackled problems previously deemed "hard" by the expert system algorithms of that time, such as backing up a tractor-trailer. Nevertheless, limited computing resources led to their dismissal for many years as mere solutions to "toy problems," even though they demonstrated the capacity to solve unforeseen challenges with limited computing resources.

9.13.3.4 Recent Popularization of Neural Computing

The renewed interest in AI and ML has recently demonstrated its practicality by training on a vast dataset compiled from over a billion websites. The multi-layered network architecture and learning principles in deep learning are fundamentally the same as those used decades ago, only scaled up a billionfold. They utilize the same supervised training and reinforcement learning models developed over 50 years ago, applied to a training dataset that is billions of times larger.

9.13.4 Computational Resources Needed for Neural Computing

This technology has become a powerful tool for delivering solutions that rival human capabilities. For instance, ChatGPT employs LLMs to identify patterns in text and generate human-like conversations. It was trained on data from a billion websites, utilizing 500 billion tokens, 175 billion parameters, and extensive weight matrices to create a complex neural network. This framework encompasses nearly a trillion variables, crucial for computation and storage, facilitating the establishment of relationships among higher-order text strings as tokens.

The computing resources are vast, encompassing speed, memory, and storage. Generating text dialogues comparable to human conversations required a month of calculations on cloud server farms that were previously accessible only to major tech companies. Although these dialogues appear related to the topic, they may lack relevance without proper context. Current AI models excel at summarizing large datasets by generalizing based on the higher-order correlation of input-output relationships; however, they do not yet possess the capability to perform logical reasoning or deduction.

9.13.4.1 Generating Credible Responses from Patterns without Reasoning

With the computing resources available in cloud server banks, neural networks can summarize large volumes of complex data, such as analyzing combinations of genetic profiles from extensive DNA datasets to identify interrelated relationships. However, these connections represent plausible relationships that may not be based on logic but rather on patterns. Therefore, AI-generated genetic matches are plausible but not necessarily valid.

9.13.4.2 Legal Criteria for Validity

Legal criteria require that the forensic analysis exceed a reasonable doubt. Concerns regarding AI-generated relationships arise because they may appear valid based on random associations instead of logical reasoning. Therefore, they do not inherently satisfy the legal standards for conviction or determining parenthood matches.

9.13.4.3 Less Restrictive Requirements for Plausible Heritage

Exploring heritage can be an enjoyable exercise in genealogical tracing. It seeks to uncover potential lineage rather than adhere to strict criteria for definitive results. Therefore, using AI to suggest possible heritage is only a recommendation, even if the lineage is hypothetical rather than guaranteed. Examining potential heritage can be a rewarding experience.

9.13.5 Bridging the Information Gaps

Unlike traditional expert systems, AI trained through ML can identify connections even when there are gaps in the training data. This capability effectively fills information voids by suggesting plausible links rather than leaving them unexplored as unknown.

9.13.5.1 Unexplained Conclusions

AI ML systems connect inputs to outputs by identifying relationships within training datasets. By examining the input-output patterns observed during training, we can draw conclusions. However, this method often lacks transparency, making it difficult to understand how outputs are generated from billions of examples without a logical explanation.

The connection may be plausible, but it lacks logic. Without understanding the reasoning behind the conclusions, precautions should verify the validity of the results. However, even without a strong rationale, the results can still support exploratory analysis and highlight specific features for human assessment without entirely dismissing them.

9.13.6 Explainable AI (XAI)

Explainable AI (XAI) research clarifies the reasoning behind AI responses by aligning response patterns with logical explanations. This approach promotes interpretable conclusions. It enhances confidence in their validity rather than leaving them unexplainable.

9.14 CASE HISTORY: ARTIFICIAL NEURAL NETWORKS

A historical overview of artificial neural networks and artificial neurons in computing is essential for understanding AI systems that exhibit human-like intelligence. Understanding these principles enhances our trust in AI-generated outcomes and deepens our comprehension of genetics and inheritance. Trainable artificial neural systems depend on established physical and mathematical computational principles rather than mystical ones.

9.14.1 Bio-Inspired Neuron Computation

The theoretical foundation for neural processing in neural networks is based on biological neurons' anatomical and physiological principles of biological neurons. Artificial neurons were inspired by the works of neuroanatomist Santiago Ramón y Cajal, neurophysiologist Sir Charles Scott Sherrington, and neuropsychologist William James. Their theories linked the fundamental functions of neurons to behaviors, establishing a connection between the mind and body.

9.14.1.1 Case Example: The McCulloch-Pitts Neuron

In 1943, Warren McCulloch and Walter Pitts[35] demonstrated that neurons can perform logical operations through threshold activation functions. A perceptron is a binary classifier that uses supervised learning rules to derive solutions to classification problems from training data without explicit instructions.

9.14.1.2 Case Example: The Perceptron Neuron

Artificial neurons (see Figure 9.7) have shown the ability to perform the logical OR operation with a threshold of 1 (see Table 9.1) and the logical AND operation with a threshold of 2 (see Table 9.2). In this model, inputs and outputs are represented as 1 (true) or 0 (false), consistent with the original Perceptron model.

9.14.1.3 Computing the Boolean Logical OR Operation with Artificial Neurons

A truth table (see Table 9.1) illustrates the connection weights that define input-output relationships. In this example, the connection weights are set to 1. By adjusting these weights, machines can learn from examples

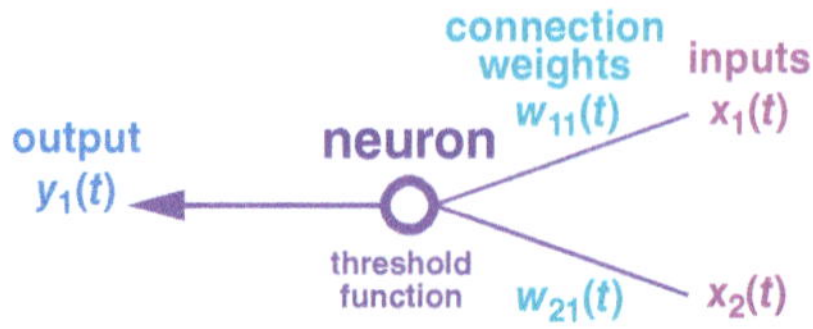

FIGURE 9.7 A diagram illustrating a McCulloch-Pitts neuron highlights the relationships between inputs, outputs, and connection weights. The connection weight calculates the weighted sum of the neuron's inputs. The threshold activation function determines the neuron's output, activating when the weighted sum reaches a specific threshold. It computes the logical OR function when the threshold is set to 1 and the logical AND function when set to 2, with both inputs and outputs represented as 1 (true) or 0 (false).

without explicit instructions on solving problems. When the threshold is set to 1, the neuron computes the Boolean logical OR operation.

9.14.1.4 Computing the Boolean Logical AND Operation with Artificial Neurons

When the threshold is set to 2, the truth table (see Table 9.2) illustrates that it computes the Boolean logical AND operation. This concept was instrumental in using artificial neurons for AI computation. They can also compute other logical NOT (negation), NOR (NOT OR), and NAND (NOT AND) operations.

9.14.1.5 Case Example: The Perceptron Machine

In 1958, Frank Rosenblatt developed a perceptron machine for image processing, enabling it to classify images through training rather than programmed instructions.[36] It could distinguish between cats and non-cats without needing explicit directions. Although it initially generated excitement in AI research, the capabilities of neural computation faced criticism due to its single-layer architecture, particularly when it could not solve problems like XOR computation.

9.14.1.6 The AI Winter

In 1969, Marvin Minsky and Seymour Papert published a book[37] critiquing the limitations of perceptrons because of their inability to solve the XOR computation. As a result, funding for neural network research dwindled for decades. The algorithmic approach was preferred over ML for addressing complex problems.

TABLE 9.1 This truth table illustrates the neural computation of the logical OR operation with a threshold of 1. The output activates when the weighted sum of inputs exceeds this threshold, with inputs and outputs represented as 1 (True) or 0 (False)

Logical OR Operation Using a Threshold of 1						
Input, $x_1(t)$	**Input, $x_1(t)$**	**Weight, $w_{12}(t)$**	**Weight, $w_{11}(t)$**	**Weighted Sum, $\Sigma = w_{11}(t) + w_{21}(t)$**	**Threshold, $\theta = 1$**	**Output, $y_1(t)$**
0	0	1	1	0	$\Sigma < \theta, y_1(t) = 0$	0
0	1	1	1	1	$\Sigma \geq \theta, y_1(t) = 1$	1
0	0	1	1	1	$\Sigma \geq \theta, y_1(t) = 1$	1
1	1	1	1	2	$\Sigma \geq \theta, y_1(t) = 1$	1

TABLE 9.2 This Truth Table Illustrates the Neural Computation of the Logical AND Operation with a Threshold of 2

Logical AND Operation Using a Threshold of 2						
Input, $x_1(t)$	**Input, $x_1(t)$**	**Weight, $w_{11}(t)$**	**Weight, $w_{11}(t)$**	**Weighted Sum, $\Sigma = w_{11}(t) + w_{21}(t)$**	**Threshold, $\theta = 2$**	**Output, $y_1(t)$**
0	0	1	1	0	$\Sigma < \theta, y_1(t) = 0$	0
0	1	1	1	1	$\Sigma < \theta, y_1(t) = 0$	0
0	0	1	1	1	$\Sigma < \theta, y_1(t) = 0$	0
1	1	1	1	2	$\Sigma \geq \theta, y_1(t) = 1$	1

9.14.2 The Dominance of Traditional Algorithmic Expert Systems

A common issue is the lack of sufficient DNA data in databases that compute matches, such as the missing genetic profile information in a genetic index database. Traditional algorithmic AI approaches have been the gold standard for expert systems in addressing these challenges for decades, even though they often struggle with straightforward problems humans can easily solve. These "hard" problems remain unresolved because no practical, explicit algorithmic solutions exist.

9.14.2.1 The "Hard" Problems in Algorithmic AI

Examples of challenging problems for traditional AI include filling in missing information in expert systems, such as addressing gaps in DNA databases to trace ancestry, until a neural network is used to supplement the missing profile information in those databases. Other challenging problems that seem obvious for humans to solve include speaker-dependent speech recognition, scale- and rotation-invariant handwriting recognition, and orientation- and viewpoint-invariant facial recognition in image analysis. These complex issues are now solvable through ML. Neural computation identifies patterns in training datasets to generalize solutions without needing algorithmic instructions.

9.14.3 Computations Using Artificial Neurons

An artificial neuron processes input signals from multiple sources to produce a single output (see Figure 9.7). It is analogous to a biological neuron, which receives thousands of synaptic inputs through its dendrites and transmits output via a single axonal pathway to other neurons. The application of artificial neurons for computation was inspired by the anatomical structure of biological neurons, which possess dendritic branches resembling tree branches.

9.14.3.1 Simultaneous Parallel Processing with Artificial Neurons

The design of neural computing uses neurons to gather inputs from thousands of other neurons, performing weighted-sum operations concurrently rather than sequentially. Each neuron operates as an independent processor, allowing it to process information simultaneously with other neurons in the network. This emphasizes the distinction between parallel processing in neural computation and serial processing in traditional digital CPUs.

9.14.3.2 Differences Between Neural and Conventional Digital Processing

Neural computing processes thousands of inputs simultaneously and in parallel, while conventional digital computing handles multiple inputs sequentially, addressing one or two at a time. This serial method creates bottlenecks at the CPU when dealing with a large number of inputs. The speed advantage in neural computing arises from processing signals in parallel instead of in series.

9.14.3.3 The Neural Processing Chip Hardware Accelerator

Neural computing can be computationally expensive for large-scale matrix multiplications involving floating-point and logical operations (see sections below). Conventional digital computing alleviates CPU bottlenecks by enhancing processing speed. However, modern AI systems utilizing large-scale LLM models analyze billions of inputs; for instance, processing 100 billion mathematical operations with a trillion parameters at 10 GHz takes 10 seconds for a single step in this context. Training a general purpose neural network requires billions of iterative steps before the system converges on a potential solution. Estimates indicate that power consumption by AI servers will exceed 20% of total power consumption in the U.S. within the next five years. Nevertheless, a smaller, special-purpose AI system designed for customized solutions will consume significantly fewer resources and necessitate less training data.

9.14.3.4 Computational Bottlenecks

To alleviate bottlenecks, neural computing chips are designed to execute matrix operations as a unified computing unit instead of sequentially on a conventional CPU. Furthermore, processing thousands of inputs necessitates a substantial memory footprint, which requires on-chip memory banks. Analog computing, similar to biological neurons, could significantly enhance speed and reduce power consumption.

9.14.3.5 Accelerating Computational Speed with Analog Signals in Neurons

Biological neurons evolved to overcome computational bottlenecks by employing analog signal processing. They process signals electrically through membrane potentials, integrating thousands of inputs as analog signals. Excitatory inputs generate positive potentials, while inhibitory inputs create negative potentials for summation. The membrane potentials aggregate synaptic inputs as graded potentials.

9.14.4 Mathematical Computation in Biological Neurons

They perform mathematical operations for addition and subtraction using both positive and negative potentials. The attenuation adjusts the weighted sum by scaling the potential amplitude along the path to the integration zone. Signals are attenuated based on the impedance of the signal path and the distance from the synapse to the axon hillock. Most signal paths from a synapse follow a dendritic branch unless the synapses are located on the soma.

9.14.4.1 Nonlinear Operations as a Threshold in Biological Neurons

The firing threshold defines the threshold function for activation based on whether the weighted sum surpasses it. The gating mechanisms of the ionic channels embedded in the membrane convert analog signals into digital signals in the form of action potentials, which are transmitted to the next neuron for processing.

9.14.4.2 Digital Signal Transmission in Biological Neurons

To maintain signal transmission integrity, neurons transmit digital pulse-coded signals encoded by action potentials over long distances along the axonal membrane. Upon reaching the axon terminal, these action potential signals are converted into chemical signals that act as neurotransmitters. This system primarily functions as a hybrid for processing electrochemical signals. By analogy, most batteries also operate as hybrid electrochemical systems.

9.14.4.3 Chemical Signal Transmission in Biological Neurons

Before transmitting signals to the next neuron, neurons convert electrical signals into chemical signals in the form of neurotransmitter packets that encode excitatory or inhibitory responses. An excitatory response increases the likelihood of the next neuron's firing, while an inhibitory response decreases this probability. This neurotransmitter release process acts as a universal mechanism for encoding both types of signals to regulate the firing of the subsequent neuron.

9.14.4.4 Transmission of Excitatory and Inhibitory Signals in Neurons

An excitatory synapse uses specific neurotransmitters as chemical signals that bind to receptors, converting these pulse codes into a positive membrane potential in the next neuron and facilitating addition. Conversely, an inhibitory synapse employs a different neurotransmitter that binds to a distinct receptor, transforming the signals into a negative potential in the subsequent neuron and enabling subtraction.

9.14.4.5 Parallel Processing in Biological Neural Networks

Each neuron processes both analog and digital signals using the same mechanisms, accelerating the overall process by concurrently managing signals among billions of neurons in the brain. This capability enables massive parallel neural computing, effectively addressing the limitations of serial computing. Although the processing speed of biological neurons is a million times slower than that of electronic computers, the brain can simultaneously compute at high speeds by processing computations across billions of neurons.

9.14.5 Threshold Processing in Biological Neurons

The gating mechanisms of ionic channels at the trigger zone initiate action potentials at the axon hillock. When the summed membrane potential exceeds this threshold, the weighted-sum analog signal is converted into a digital signal as an action potential.

9.14.5.1 Digital Signal Processing in Biological Neurons

The action potential is a digital pulse signal produced by a neuron. It is characterized by a consistent amplitude and width, which encode the neural response as a time series of action potentials. The neuron transmits these digital signals along the axon as outputs for the next neuron to process.

9.14.5.2 Hybrid Chemical and Electrical Signal Processing in Biological Neurons

The signal is converted into a chemical signal through neurotransmitters released at the synapse. This chemical signal can be either excitatory or inhibitory, depending on whether it transmits a go or no-go signal to continue or block communication. The processing advances to the next layers of neurons, resulting in collective rather than individual computation by neurons.

9.14.5.3 Voltage-Gated Channels as Differential Equation Solvers for Biological Neurons

The firing threshold is regulated by a set of ionic channels embedded in the membrane. Each ionic channel in a neuron has voltage gates that govern a series of differential equations related to voltage activation. A typical neuron contains sodium (Na) channels and potassium (K) channels that regulate this threshold.[38] As Alan Hodgkin and Andrew Huxley described, these thresholds were controlled by gating channels and rectifiers between 1948 and 1952.[39–45] Each neuron effectively processes a set of differential equations through analog signal processing, driven by the regulation of threshold voltage. Hodgkin and Huxley were awarded the Nobel Prize in 1963 for their contributions to neural computation, particularly for discovering the gating properties of these ionic channels[46] before their existence was confirmed.

9.14.5.4 Neural Computation with Voltage-Gated Channels

Erwin Neher and Bert Sakmann later employed the patch-clamp technique to confirm the existence of gating currents generated by these ionic channels. They were also awarded the Nobel Prize in 1991.[47] This exemplifies a case of discovering biological neural computations that inspired processing with artificial neurons.

9.14.5.5 Synaptic Relays in Biological Neurons

The synapse is where electrical signals travel between neurons across a narrow gap. Neurotransmitters transport these signals across the gap, converting them into a voltage in the following neuron. Biological neurons process information using electrical signals, relaying results to the next neuron through a complex transformation into chemical signals before returning to electrical signals for further processing.

Each connection influences firing, with certain connections having a greater impact due to their synaptic strength and proximity to the soma (cell body). A neuron activates when the total synaptic inputs at the axon hillock exceed a threshold, generating action potentials (nerve impulses).

9.14.6 Mathematical Weighted Sum in Biological Neurons

A neuron acts as an adder, summing inputs from earlier layers in the neural network. When this computational process occurs for all neurons in a network, signals are processed progressively along the pathways of each neuron in parallel. In other words, as conventional computers do, neural computation distributes processing across all neurons without depending on a single CPU.

9.14.7 Neurons as a Voting System: Counting Votes from Synaptic Inputs

Each connection to a neuron carries a weight that influences its responses. When all weights are equal, the neuron receives unbiased inputs. It operates like a voting system, gathering input votes to determine the output based on a threshold. This threshold signals whether the votes are sufficient to declare a winner, similar to a simple majority or a supermajority in decision-making. The neuron activates its output when the total count exceeds the threshold, reflecting the outcome of the voting process.

9.14.7.1 The Use of Input Bias in Neural Computation

In a fair election, each vote is counted equally. This means all connection weights are unbiased and treated the same when calculating the total. However, if some connection weights are greater than others, those votes hold more significance. Biological neurons manipulate biases in vote counting to adjust the importance of specific inputs for learning and survival.

9.14.7.2 Adjusting Relevant Inputs via Connection Weights

To survive, one must focus on essential stimuli and ignore irrelevant ones. If all stimuli were equally significant, sensory overload could overwhelm an individual. Therefore, learning emphasizes inputs related to survival while filtering out unimportant ones. By adjusting synaptic weights based on learning principles, the system homes in on crucial stimuli and dismisses those that provide less relevant information.

9.14.8 The "Use It or Lose It" Principle of Synaptic Plasticity

Biological neurons utilize Hebbian learning to modify synaptic weights. This activity-dependent rule reinforces connections when neural inputs are activated simultaneously. It reflects the "use it or lose it" principle. Cognitive decline in Alzheimer's patients is mitigated when they participate in mental activities.

9.14.8.1 The Auto-Associative Hebbian Learning Rule

The Hebbian learning rule is an auto-associative learning mechanism employed for unsupervised learning, meaning it does not require a "teacher" to train the system. It learns to associate relevant inputs that are activated simultaneously. The closer the temporal proximity, the stronger the association.

9.14.8.2 Learning Rules for Modifying Connection Weights

The learning rule is simple: strengthen important connections and diminish less significant ones. By modifying connection weights, the system prioritizes responses to critical stimuli. Over time, this process slowly enhances suitable responses by emphasizing positive feedback while ignoring negative feedback. As a result, adjusting connection weights to favor selected inputs improves our overall effectiveness.

9.14.8.3 The Choice of Feedback for Adjusting Connection Weights

Determining which stimuli to prioritize depends on the value of the feedback. Similarly, neurons select specific types of feedback to guide the learning rules for adjusting their connection weights. Because some feedback is more effective than others, this decision ultimately influences the results of appropriate responses.

9.14.9 Principles of Neural Computation: The Weighted Sum

If $x_i(t)$ represents the i-th input to the neuron at time t, and $w_i(t)$ denotes the connection weights (see Figures 9.7 and 9.9), then the weighted sum $s(t)$ is computed as follows:

$$s(t) = w_1(t)x_1(t) + \cdots + w_i(t)x_i(t) \cdots + w_n(t)x_n(t) = \sum_{i=1}^{i=n} w_i(t)x_i(t) \tag{9.2}$$

9.14.9.1 The Vector Representation of Neural Inputs

This indicates that an input vector $\vec{X}(t)$ represents all n inputs to the neuron:

$$\vec{X}(t) = \begin{bmatrix} x_1(t) \\ \vdots \\ x_n(t) \end{bmatrix} \tag{9.3}$$

9.14.10 Principles of Matrix Multiplication in a Network Layer

In a simple layer of m input neurons, each with n inputs and $w_{ij}(t)$ representing the connection weights of the j-th neuron, the weighted sum is computed using the following matrix equation:

$$\begin{bmatrix} s_1(t) \\ \vdots \\ s_m(t) \end{bmatrix} = \begin{bmatrix} w_{11}(t) & \cdots & w_{n1}(t) \\ \vdots & \ddots & \vdots \\ w_{1m}(t) & \cdots & w_{nm}(t) \end{bmatrix} \begin{bmatrix} x_1(t) \\ \vdots \\ x_n(t) \end{bmatrix} \tag{9.4}$$

9.14.11 Principles of Internal Representation by Weight Matrices

Represent the above equation using matrix notation as follows:

$$\vec{S}(t) = \vec{W}(t) \cdot \vec{X}(t) \tag{9.5}$$

The equation above illustrates the matrix multiplication performed by each neuron. The matrix $\vec{W}(t)$ represents the connection weights $w_{ij}(t)$ from the i-th neuron to the j-th neuron during the matrix multiplication process:

$$\vec{W}(t) = \begin{bmatrix} w_{11}(t) & \cdots & w_{n1}(t) \\ \vdots & \ddots & \vdots \\ w_{1m}(t) & \cdots & w_{nm}(t) \end{bmatrix} \tag{9.6}$$

This matrix computation occurs before a threshold function is applied to determine the output. If the weighted sum exceeds this threshold, the neuron activates its output and transmits activation information to the next neuron for further processing. If it does not exceed the threshold, the neuron remains inactive until the next time step, repeating the same process.

9.14.12 The Size Principle of Neural Networks

A biological neuron typically receives about 10,000 to 100,000 synaptic inputs from other neurons and performs weighted-sum calculations in a single step that takes less than a millisecond. The human brain contains approximately 100 billion neurons, each with an average of 10,000 to 100,000 inputs, and its computational requirements are comparable to those of current AI systems.

9.14.12.1 The One-Step Processing Principle in Neural Computation

Processing speeds improve when all neurons carry out these mathematical operations simultaneously instead of one node at a time. In today's AI networks, each artificial neuron can have over a trillion parameters to update at each time step. The bottleneck arises from the serial processing performed on conventional multi-core CPUs.

9.14.12.2 Bio-inspired Neuron-Based Weighted Sum Computation

Biological neurons inspire neural processing in artificial neurons. Each biological neuron has thousands of synaptic inputs on its dendritic branches and soma, resembling the leaves of a tree branch. It integrates synaptic potentials at the axon hillock to determine whether to fire, but only if the summed voltage reaches a certain threshold. It enhances matrix computation by summing voltages as analog signals from distant synaptic inputs, weighted by the attenuation along the dendritic tree branch.

9.14.12.3 Computing Uniform Contributions from Inputs

If all the weights are equal, the system simplifies to a straightforward sum of the inputs. Each input contributes equally, resulting in an unbiased system. Neural computation becomes a simple "adder" operation, similar to how a computer performs addition within the CPU. The computational processes in both digital and neural computing are analogous, illustrating the similarities between these two mechanized computations.

9.14.12.4 Principles for Adjusting Connection Weights According to Input Contributions

If each connection weight changes individually, the neuron adjusts the contribution of each input according to its updated weight. It gradually adapts to the significance of each input's contribution, leading to learning through the reorganization of its connections with other neurons.

9.14.13 Principles of Computation for Correlation Functions

Higher connection weights significantly enhance the weighted sum, resulting in a positive correlation. Conversely, negative weights diminish the sum, leading to a negative correlation. Time-delayed temporal processing computes the correlation function for a time series of pulse-coded action potentials.[48] The temporal correlation functions generated by biological neurons emerge from their processing of spike trains, which creates the association between input signals represented by an internal correlation matrix.[49–52] Neurons perform temporal integration of synaptic inputs for processing.[53, 54] This process enables neurons to learn to generate appropriate outputs by recognizing higher-order correlations for each connection, even if these are indirectly related.

9.14.13.1 Principles of Hebbian Learning Rule

As explained earlier, the learning rule gradually adjusts weights, allowing the system to respond to relevant inputs for the desired output. The adjustment rule is straightforward: strengthen the connection by increasing the weight when the input is relevant and decrease it when it is not. This process modifies the influence of relevant inputs on neuron firing when activated. The connection is weakened for irrelevant inputs by lowering the weight when not activated.

9.14.13.2 Principles of Emphasizing Contributions of Relevant Inputs by Weights

The neuron adjusts its connection weights over time, amplifying relevant inputs and diminishing irrelevant ones. Eventually, its weights stabilize to reflect the importance of these connections. Only relevant inputs affect firing based on their weighted sums, while irrelevant inputs have minimal influence.

9.14.14 Principle of the Output Activation Hard Threshold Step Function

The threshold activation function acts as a step function for an artificial neuron (see Figure 9.8). It is discrete and discontinuous, defined by a precise threshold. At this threshold, differentiation becomes undefined. As a nonlinear function, it cannot be traced back from the output to the original inputs. The output activation changes abruptly as a step function. It is 1 when the weighted sum exceeds the threshold; otherwise, it is 0:

$$y_j(t) = \begin{cases} 1, \text{if } \sum_{i=1}^{i=n} w_{ij}(t)x_i(t) \geq \theta \\ 0, \text{if } \sum_{i=1}^{i=n} w_{ij}(t)x_i(t) < \theta \end{cases} \tag{9.7}$$

9.14.14.1 The Sigmoid Soft Threshold Activation Function

Other threshold functions, such as the sigmoidal function (see Figure 9.8), provide a gradual change in output activation rather than a discrete step change. The threshold activation function is smooth and continuous, acting as a soft threshold. It is a nonlinear function that cannot be reversed from the output to the original inputs.

9.14.14.2 Principles for Collapsing Linear Threshold Functions into a Single-Layer Network

A linear threshold (see Figure 9.8) represents a continuous function equivalent to no threshold. As a linear function, it does not change the output activation produced by the weighted sum. The activation function equals the weighted sum of the inputs. In such cases, the series of matrix

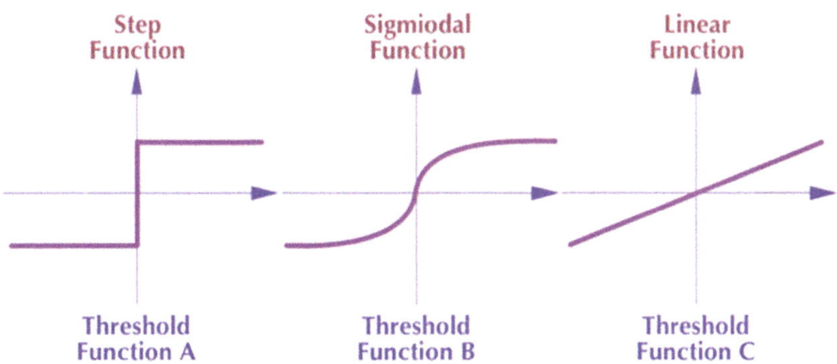

FIGURE 9.8 Graphs illustrating various types of threshold functions. A step function threshold is a discrete, discontinuous function that indicates a hard threshold. A sigmoidal function threshold is a continuous function that signifies a soft threshold. A linear function threshold effectively represents the absence of a threshold.

multiplications can be simplified into a single equivalent matrix of a single-layer network, which does not perform any significant processing.

9.14.15 Principles of Multi-Layer Neural Networks

A multi-layer neural network consists of hidden layers situated between the input and output layers (see Figure 9.9). Each hidden layer processes and extracts essential information to address more complex tasks. Each nonlinear layer abstracts information related to the input-output relationships, facilitating effective feature extraction. As the network learns, it gradually adjusts its internal representation of these relationships. The learning rule modifies the connection weights according to the significance of these input-output relationships. Since these relationships are expressed as matrices, they encapsulate the internal model representation. Each hidden layer captures a unique representation, extracting different connections to provide higher-order input-output relationships.

9.14.15.1 Principles of Matrix Multiplication in Neural Computation

Representing the weighted sum of inputs with a matrix results in an output activation as follows:

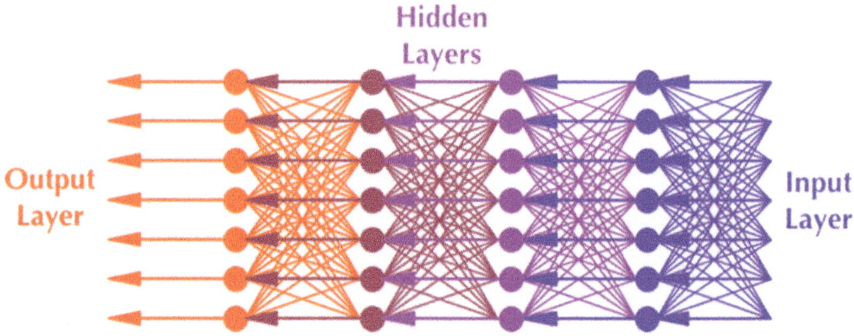

FIGURE 9.9 A multi-layer neural network illustrating the different layers and their input-output relationships. The hidden layers between the input and output layers capture complex connections among the inputs and outputs.

$$y_j(t) = \begin{cases} 1, \text{if} \begin{bmatrix} w_{11}(t) & \cdots & w_{n1}(t) \\ \vdots & \ddots & \vdots \\ w_{1m}(t) & \cdots & w_{nm}(t) \end{bmatrix} \begin{bmatrix} x_1(t) \\ \vdots \\ x_n(t) \end{bmatrix} \geq \theta \\ 0, \text{if} \begin{bmatrix} w_{11}(t) & \cdots & w_{n1}(t) \\ \vdots & \ddots & \vdots \\ w_{1m}(t) & \cdots & w_{nm}(t) \end{bmatrix} \begin{bmatrix} x_1(t) \\ \vdots \\ x_n(t) \end{bmatrix} < \theta \end{cases} \tag{9.8}$$

9.14.15.2 Principles of Computed Output Activation Functions

By utilizing matrix notation to simplify the matrices, the output activation function can be expressed by the following equation:

$$\vec{Y} = \begin{cases} 1, & \text{if } \vec{W} \cdot \vec{X} \geq \theta \\ 0, & \text{if } \vec{W} \cdot \vec{X} < \theta \end{cases} \tag{9.9}$$

9.14.15.3 Principles of Computation with Irreplaceable Hidden Layers

By cascading the outputs of one layer to the next, the neurons in the intermediate layers can use the output of each neuron as input. The intermediate layer plays a crucial role in neural computation and signal processing. It addresses the limitations of models like the perceptron, which cannot solve the XOR problem due to its single-layer structure. Hidden layers facilitate nonlinear processing through the threshold function; without them, computations may revert to a single layer.

9.14.15.4 Principles of Nonlinear Neural Network Computation

The threshold function, a binary step function that performs Boolean logic, is mathematically nonlinear. This nonlinearity prevents the series of matrix multiplications from simplifying into a single equivalent multiplication. If it were otherwise, the computations of a multi-layer network would be reducible to those of a single-layer neural network's matrix multiplication.

9.14.15.5 The Essential Processing by Intermediate Layers

The hidden layers are essential for computations in neural networks. Intermediate interneurons would be unnecessary if a single-layer equivalent could replace the network. Due to nonlinear functions, these layers compute critical behaviors that simpler networks cannot replicate.

9.14.15.6 The Essential Processing of Multi-layer Neural Networks

Logical inference and causal reasoning in network connections stem from nonlinear neural processing. These processes are irreversible due to the noncommutativity of matrix multiplication, which renders it impossible to trace neural networks back to the origins of consciousness's abilities. This distinction distinguishes AI from traditional preprogrammed algorithmic approaches to autonomously solving intellectual processing functions.

9.14.15.7 Principles of Methodologies for Adjusting Internal Representations

The principle involves learning rules to adjust connection weights based on network feedback to minimize errors. These five processes will impact the network's performance:

- The quality and quantity of the dataset for training a network;
- The learning rules for updating the connection weights between neurons;
- The selection of relevant feedback for determining the above updates;
- The network criteria for improving the outcomes of the network outputs;
- The methodology for avoiding convergence into a non-optimal solution.

9.14.16 Principles of the Optimization Process in AI Training

ML in AI models is a mathematical, data-driven methodology that optimizes solutions by minimizing errors to align with expected results. It employs a trial-and-error approach to learn from mistakes. By reducing discrepancies between the system's output and the anticipated results in each iteration, the system typically converges on a solution that achieves the desired outcomes.

9.14.16.1 Principles of the Generalization Process in AI

Generalizing a solution requires considerable time and numerous training examples to determine expected outcomes by establishing extensive relevant connections. This process fosters an internal representation that enables generalization from a training dataset, even if the input query is not part of that dataset. The key lies in the internal representation created by a network of neural connections that establishes higher-order correlations in the input-output relationships.

9.14.17 Evaluation of AI Network Performance

Understanding AI model training is crucial for selecting a model for genealogical analysis. A network's performance hinges on the learning rules used during training and its connectivity to determine relevance. This comprehension enables an objective evaluation rather than relying solely on vendor marketing. Below is a summary of AI methodologies for assessing performance.

9.14.17.1 Principles of Minimizing Error Methods

The backprop rule reduces errors by progressively propagating them from the output layer to the hidden layers. The error is calculated by comparing the network's outputs with the desired outputs and is then used to update the connection weights for each neuron. This method minimizes errors through gradient descent and was developed by Cauchy in 1847.[55]

9.14.17.2 Principles of Backprop Networks Commonly Used in AI

Backprop, short for "back-propagating error correction," was coined by David Rumelhart, Geoffrey Hinton, and Ronald Williams.[56, 57] Numerous predecessors contributed to the application of backprop in neural networks. Shunichi Amari suggested training multi-layer perceptrons (MLPs) in 1967[58] with end-to-end connections using stochastic gradient descent (SGD).[59] The concept of backpropagation was introduced by Rosenblatt in 1962. The backprop model gained popularity from the book "Parallel Distributed Processing."

9.14.17.3 Limitations of the Backprop Machine Learning Model

The backprop model faces various computational challenges. A supervised learning model requires a "teacher" to provide correct answers for training. The model adapts to the training set as guided by the trainer. Training a supervised learning AI model with human input on correct responses is labor-intensive; for instance, it involves teaching the system to differentiate between male and female faces in facial recognition. In contrast, other intelligent systems utilize self-learning through the unsupervised learning paradigm to categorize independently without any instructions via auto-association.

It uses a gradient descent algorithm to find an optimal solution, but may get stuck in a local minimum (see below). Further methods are needed for the system to ascend before descending to a better solution.

9.14.18 The Gradient Descent Method for Error Minimization

One method for solving the error minimization problem is gradient descent, which follows the gradient downhill (see Figure 9.10). However, the network may converge to a solution and become trapped in a local minimum, representing a non-optimal solution. Figures 9.10 illustrates the topology of the solution space.

The discovered solution may be valid but might not be optimal. Often, one must ascend before descending into another minimum, which requires using alternative methods to counter the gradient descent algorithm. However, achieving a global minimum is not guaranteed if the optimal solution is unknown beforehand.

9.14.18.1 Methods for Escaping Local Minima

Various analytical methods aim to escape local minima, often using stochastic and probabilistic output activation functions. Semi-random fluctuations help ascend against the gradient toward an optimal solution. This approach generates random outputs to explore nearby solution spaces while moving against the downward gradient. It employs stochastic techniques rather than deterministic ones. Stochastic processing is probabilistic, solving problems differently from conventional deterministic approaches that specify exact solutions without variations. Some stochastic solutions even tunnel through the hill to escape local minima.

9.14.19 Case Example: The Boltzmann Machine

Hinton employs the Boltzmann machine, drawing on the analogy from statistical physics regarding energy to define the exploration function in a search through the solution space. Random thermal fluctuations enable the system to acquire enough energy to escape from local minima.

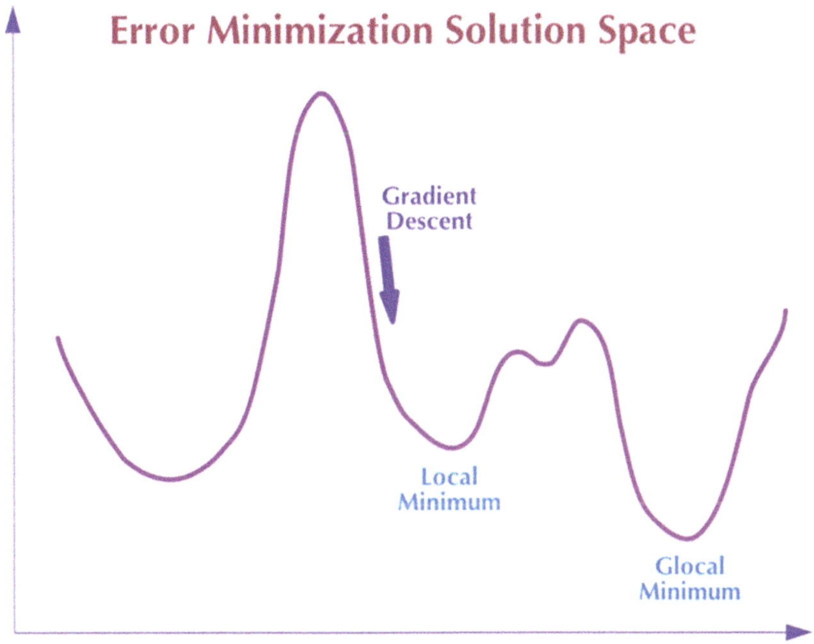

FIGURE 9.10 A graph illustrating gradient descent for error minimization highlights the risk of encountering local minima. Escaping these non-optimal solutions often requires counterintuitive approaches, such as moving against the gradient uphill before converging on a more optimal solution. It is often impossible to determine the location of the global minimum or to confirm whether the best solution has been found without prior knowledge of that optimal solution.

The temperature of the system's energy dictates the extent of exploration. Instead of the supervised learning model of backprop, the Boltzmann machine uses unsupervised learning for auto-association. Hinton received the Nobel Prize in 2024 for utilizing statistical physics energy functions to optimize solutions in the Boltzmann machine.[60]

9.14.20 Principles of Supervised Learning Training Method

A drawback of backprop is its dependency on supervised learning for training. This approach necessitates prior knowledge of the correct answers, which is unlike unsupervised learning, where such knowledge is not required. Moreover, training involves direct interaction and demands human guidance.

Human input is crucial to minimizing AI errors when training self-driving cars to recognize pedestrians and traffic lights. This process is labor-intensive and time-consuming. While they may seem automated, supervised learning AI models necessitate significant human training before arriving at an effective solution.

9.14.21 The Unsupervised Learning Training Method

The unsupervised learning model discovers solutions by forming associations between input-output pairs. Donald Hebb initially proposed the biological mechanisms for synaptic plasticity, suggesting how neurons modify synaptic strengths. The Hebbian learning rule states that simultaneous activation strengthens connections while inactivity weakens them. This process allows neurons to adjust weights without an external "teacher." The learning rule essentially establishes correlation functions autonomously, reflecting the "use it or lose it" principle introduced earlier.

9.15 SUMMARY

Genealogical and forensic analyses of genetic data are illustrated with case examples to introduce the core principles of probabilistic inheritance and analytical techniques for determining ancestral lineages and identity matches. Traditional algorithmic approaches depend on expert systems that utilize defined statistical inference methodologies to address these issues analytically; however, they cannot resolve unexpected scenarios and manage missing genetic information.

Alternative AI methodologies utilize ML to demonstrate their benefits and practical limitations. Key ML principles are summarized to provide an informed evaluation of AI tools' performance without suggesting they possess magical abilities in solving genealogical challenges. The AI approach requires extensive training on large datasets to identify suitable solutions and learns from examples to summarize data through input-output correlation. Consequently, it can bridge gaps in genealogical databases, although these outcomes may not always be logically sound.

Until AI improves its reasoning capabilities, caution is advised when drawing conclusions. However, these systems effectively automate the cataloging of genealogical databases from various records and can generate automated responses to inquiries using natural language processing to answer questions from everyday users. Understanding the mechanics of AI ML will assist in a thorough evaluation of the performance of different AI models for specific genealogical applications.

REFERENCES

(1) Bush, W. S. Genome-Wide Association Studies. In *Encyclopedia of Bioinformatics and Computational Biology*; Ranganathan, S., Gribskov, M., Nakai, K., Schönbach, C., Eds.; Academic Press: Oxford, 2019; pp 235–241. https://doi.org/10.1016/B978-0-12-809633-8.20232-X

(2) Ala-Korpela, M.; Kangas, A. J.; Inouye, M. Genome-Wide Association Studies and Systems Biology: Together at Last. *Trends in Genetics* 2011, *27* (12), 493–498. https://doi.org/10.1016/j.tig.2011.09.002

(3) Figueroa, C. J.; Tang, Y.-W.; Taur, Y. Principles and Applications of Genomic Diagnostic Techniques. In *Molecular Medical Microbiology*; Elsevier, 2015; pp 381–397. https://doi.org/10.1016/B978-0-12-397169-2.00022-6

(4) Mansur, Y. A.; Rojano, E.; Ranea, J. A. G.; Perkins, J. R. Chapter 7 - Analyzing the Effects of Genetic Variation in Noncoding Genomic Regions. In *Precision Medicine*; Deigner, H.-P., Kohl, M., Eds.; Academic Press, 2018; pp 119–144. https://doi.org/10.1016/B978-0-12-805364-5.00007-X

(5) Schifferdecker, K.; Richey, N. Artificial Intelligence and Genealogy: Promise and Problems. *Kentucky Libraries* 2024, *88*(3). https://openurl.ebsco.com/EPDB%3Agcd%3A14%3A32190126/detailv2?sid=ebsco%3Aplink%3Ascholar&id=ebsco%3Agcd%3A180828763

(6) Novroski, N. M.; Cihlar, J. C. Evolution of Single-nucleotide Polymorphism Use in Forensic Genetics. *Wiley Interdisciplinary Reviews: Forensic Science* 2022, *4* (6), e1459. https://doi.org/10.1002/wfs2.1459

(7) Bergström, A.; Stanton, D. W. G.; Taron, U. H.; Frantz, L.; Sinding, M.-H. S.; Ersmark, E.; Pfrengle, S.; Cassatt-Johnstone, M.; Lebrasseur, O.; Girdland-Flink, L.; Fernandes, D. M.; Ollivier, M.; Speidel, L.; Gopalakrishnan, S.; Westbury, M. V.; Ramos-Madrigal, J.; Feuerborn, T. R.; Reiter, E.; Gretzinger, J.; Münzel, S. C.; Swali, P.; Conard, N. J.; Carøe, C.; Haile, J.; Linderholm, A.; Androsov, S.; Barnes, I.; Baumann, C.; Benecke, N.; Bocherens, H.; Brace, S.; Carden, R. F.; Drucker, D. G.; Fedorov, S.; Gasparik, M.; Germonpré, M.; Grigoriev, S.; Groves, P.; Hertwig, S. T.; Ivanova, V. V.; Janssens, L.; Jennings, R. P.; Kasparov, A. K.; Kirillova, I. V.; Kurmaniyazov, I.; Kuzmin, Y. V.; Kosintsev, P. A.; Lázničková-Galetová, M.; Leduc, C.; Nikolskiy, P.; Nussbaumer, M.; O'Drisceoil, C.; Orlando, L.; Outram, A.; Pavlova, E. Y.; Perri, A. R.; Pilot, M.; Pitulko, V. V.; Plotnikov, V. V.; Protopopov, A. V.; Rehazek, A.; Sablin, M.; Seguin-Orlando, A.;

Storå, J.; Verjux, C.; Zaibert, V. F.; Zazula, G.; Crombé, P.; Hansen, A. J.; Willerslev, E.; Leonard, J. A.; Götherström, A.; Pinhasi, R.; Schuenemann, V. J.; Hofreiter, M.; Gilbert, M. T. P.; Shapiro, B.; Larson, G.; Krause, J.; Dalén, L.; Skoglund, P. Grey Wolf Genomic History Reveals a Dual Ancestry of Dogs. *Nature* 2022, *607* (7918), 313–320. https://doi.org/10.1038/s41586-022-04824-9

(8) Gao, S.; Li, B.; Mao, L.; Wang, W.; Zou, D.; Zheng, J.; Zhou, M.; Yu, S.; Zheng, F.; Yin, Y.; Liu, S. Q.; Yang, H.; Wang, H. A Theoretical Base for Non-Invasive Prenatal Paternity Testing. *Forensic Science International* 2023, *346*, 111649. https://doi.org/10.1016/j.forsciint.2023.111649

(9) Sachidanandam, R.; Weissman, D.; Schmidt, S. C.; Kakol, J. M.; Stein, L. D.; Marth, G.; Sherry, S.; Mullikin, J. C.; Mortimore, B. J.; Willey, D. L.; Hunt, S. E.; Cole, C. G.; Coggill, P. C.; Rice, C. M.; Ning, Z.; Rogers, J.; Bentley, D. R.; Kwok, P. Y.; Mardis, E. R.; Yeh, R. T.; Schultz, B.; Cook, L.; Davenport, R.; Dante, M.; Fulton, L.; Hillier, L.; Waterston, R. H.; McPherson, J. D.; Gilman, B.; Schaffner, S.; Van Etten, W. J.; Reich, D.; Higgins, J.; Daly, M. J.; Blumenstiel, B.; Baldwin, J.; Stange-Thomann, N.; Zody, M. C.; Linton, L.; Lander, E. S.; Altshuler, D. A Map of Human Genome Sequence Variation Containing 1.42 Million Single Nucleotide Polymorphisms. *Nature* 2001, *409* (6822), 928–933. https://doi.org/10.1038/35057149

(10) Fan, H.; Chu, J.-Y. A Brief Review of Short Tandem Repeat Mutation. *Genomics, Proteomics & Bioinformatics* 2007, *5* (1), 7–14. https://doi.org/10.1016/S1672-0229(07)60009-6

(11) Bustos, B. I.; Billingsley, K.; Blauwendraat, C.; Gibbs, J. R.; Gan-Or, Z.; Krainc, D.; Singleton, A. B.; Lubbe, S. J. Genome-Wide Contribution of Common Short-Tandem Repeats to Parkinson's Disease Genetic Risk. *Brain* 2023, *146* (1), 65–74. https://doi.org/10.1093/brain/awac301

(12) Nikolaidis, A.; Gray, J. R. ADHD and the DRD4 Exon III 7-Repeat Polymorphism: An International Meta-Analysis. *Social Cognitive and Affective Neuroscience* 2010, *5* (2–3), 188–193. https://doi.org/10.1093/scan/nsp049

(13) Wang, E.; Ding, Y.-C.; Flodman, P.; Kidd, J. R.; Kidd, K. K.; Grady, D. L.; Ryder, O. A.; Spence, M. A.; Swanson, J. M.; Moyzis, R. K. The Genetic Architecture of Selection at the Human Dopamine Receptor D4 (DRD4) Gene Locus. *American Journal of Human Genetics* 2004, *74* (5), 931–944. https://doi.org/10.1086/420854

(14) Ding, Y.-C.; Chi, H.-C.; Grady, D. L.; Morishima, A.; Kidd, J. R.; Kidd, K. K.; Flodman, P.; Spence, M. A.; Schuck, S.; Swanson, J. M.; Zhang, Y.-P.; Moyzis, R. K. Evidence of Positive Selection Acting at the Human Dopamine Receptor D4 Gene Locus. *Proceedings of the National Academy of Sciences of the United States of America* 2002, *99* (1), 309–314. https://doi.org/10.1073/pnas.012464099

(15) Slatkin, M. Linkage Disequilibrium--Understanding the Evolutionary Past and Mapping the Medical Future. *Nature Reviews. Genetics* 2008, *9* (6), 477–485. https://doi.org/10.1038/nrg2361
(16) Kreitman, M. Methods to Detect Selection in Populations with Applications to the Human. *Annual Review of Genomics and Human Genetics* 2000, *1*, 539–559. https://doi.org/10.1146/annurev.genom.1.1.539
(17) Tajima, F. Statistical Method for Testing the Neutral Mutation Hypothesis by DNA Polymorphism. *Genetics* 1989, *123* (3), 585–595. https://doi.org/10.1093/genetics/123.3.585
(18) El-Fishawy, P. Common Disease-Common Variant Hypothesis. In *Encyclopedia of Autism Spectrum Disorders*; Volkmar, F. R., Ed.; Springer New York: New York, NY, 2013; pp 719–720. https://doi.org/10.1007/978-1-4419-1698-3_1998
(19) Slatkin, M.; Rannala, B. Estimating Allele Age. *Annual Review of Genomics and Human Genetics* 2000, *1*, 225–249. https://doi.org/10.1146/annurev.genom.1.1.225
(20) Bonvicini, C.; Faraone, S. V.; Scassellati, C. Attention-Deficit Hyperactivity Disorder in Adults: A Systematic Review and Meta-Analysis of Genetic, Pharmacogenetic and Biochemical Studies. *Molecular Psychiatry* 2016, *21* (7), 872–884. https://doi.org/10.1038/mp.2016.74
(21) Kidd, K. K.; Podini, D. A Brief Introduction to Microhaplotypes and Their Uses. *Medical Research Archives* 2024, *12* (3). https://doi.org/10.18103/mra.v12i3.5142
(22) Norrgard, K. Forensics, DNA Fingerprinting, and CODIS. *Nature Education* 2008, *1* (1), 35. https://acces.ens-lyon.fr/acces/thematiques/evolution/accompagnement-pedagogique/accompagnement-au-lycee/premiere-2019/transmission-variation-et-expression-du-patrimoine-genetique/l2019histoire-humaine-lue-dans-son-genome/empreinte-genetique/forensics-dna
(23) Osborn-Gustavson, A. E.; McMahon, T.; Josserand, M.; Spamer, B. J. The Utilization of Databases for the Identification of Human Remains. In *New Perspectives in Forensic Human Skeletal Identification*; Elsevier, 2018; pp 129–139. https://doi.org/10.1016/B978-0-12-805429-1.00012-0
(24) Reilly, P. Legal and Public Policy Issues in DNA Forensics. *Nature Reviews. Genetics* 2001, *2* (4), 313–317. https://doi.org/10.1038/35066091
(25) Weir, B. S. The Rarity of DNA Profiles. *The Annals of Applied Statistics* 2007, *1* (2), 358. https://doi.org/10.1214/07-AOAS128
(26) Kircher, M.; Stenzel, U.; Kelso, J. Improved Base Calling for the Illumina Genome Analyzer Using Machine Learning Strategies. *Genome Biology* 2009, *10* (8), R83. https://doi.org/10.1186/gb-2009-10-8-r83

(27) Turing, A. M. On Computable Numbers, with an Application to the Entscheidungsproblem. A Correction. *Proceedings of the London Mathematical Society* 1938, 2 (1), 544–546.
(28) Turing, A. M. On Computable Numbers, with an Application to the Entscheidungsproblem. *Journal of Mathematics* 1936, *58* (345–363), 5. https://doi.org/10.1112/PLMS/S2-42.1.230
(29) Turing, A. Computing Engine (1947). *The Essential Turing* 2004, 362. https://www.google.com/books/edition/The_Essential_Turing/dSUTDAAAQBAJ?hl=en&gbpv=1
(30) von Neumann, J. First Draft of a Report on the EDVAC. *IEEE Annals of the History of Computing* 1993, *15* (4), 27–75. https://doi.org/10.1109/85.238389
(31) Turing, A. M. I.—Computing Machinery and Intelligence. *Mind* 1950, *LIX* (236), 433–460. https://doi.org/10.1093/mind/LIX.236.433
(32) Imperial War Museums. *How Alan Turing Cracked the Enigma Code*. Imperial War Museums. https://www.iwm.org.uk/history/how-alan-turing-cracked-the-enigma-code
(33) Orders and Medals. UK Honours System. https://honours.cabinetoffice.gov.uk/about/orders-and-medals
(34) The Nobel Foundation. NobelPrize.org. https://www.nobelprize.org/the-nobel-prize-organisation/the-nobel-foundation
(35) McCulloch, W. S.; Pitts, W. A Logical Calculus of the Ideas Immanent in Nervous Activity. *The Bulletin of Mathematical Biophysics* 1943, *5*, 115–133. https://link.springer.com/article/10.1007/BF02478259
(36) Rosenblatt, F. The Perceptron: A Probabilistic Model for Information Storage and Organization in the Brain. *Psychological Review* 1958, *65* (6), 386–408. https://doi.org/10.1037/h0042519
(37) Minsky, M.; Papert, S. An Introduction to Computational Geometry. *Cambridge Tiass., HIT* 1969, *479* (480), 104. https://leon.bottou.org/publications/pdf/perceptrons-2017.pdf
(38) Catterall, W. A.; Raman, I. M.; Robinson, H. P. C.; Sejnowski, T. J.; Paulsen, O. The Hodgkin-Huxley Heritage: From Channels to Circuits. *The Journal of Neuroscience* 2012, *32* (41), 14064–14073. https://doi.org/10.1523/JNEUROSCI.3403-12.2012
(39) Hodgkin, A. L.; Huxley, A. F.; Katz, B. Measurement of Current-Voltage Relations in the Membrane of the Giant Axon of Loligo. *The Journal of Physiology* 1952, *116* (4), 424–448. https://doi.org/10.1113/jphysiol.1952.sp004716
(40) Armstrong, C. M.; Hollingworth, S. Na(+) and K(+) Channels: History and Structure. *Biophysical Journal* 2021, *120* (5), 756–763. https://doi.org/10.1016/j.bpj.2021.01.013
(41) Hodgkin, A. L. The Local Electric Changes Associated with Repetitive Action in a Non-Medullated Axon. *The Journal of Physiology* 1948, *107* (2), 165–181. https://doi.org/10.1113/jphysiol.1948.sp004260

(42) Hodgkin, A. L.; Huxley, A. F. A Quantitative Description of Membrane Current and Its Application to Conduction and Excitation in Nerve. *The Journal of Physiology* 1952, *117* (4), 500–544. https://doi.org/10.1113/jphysiol.1952.sp004764
(43) Hodgkin, A. L.; Huxley, A. F. Currents Carried by Sodium and Potassium Ions through the Membrane of the Giant Axon of Loligo. *The Journal of Physiology* 1952, *116* (4), 449–472. https://doi.org/10.1113/jphysiol.1952.sp004717
(44) Hodgkin, A. L.; Huxley, A. F. The Components of Membrane Conductance in the Giant Axon of Loligo. *The Journal of Physiology* 1952, *116* (4), 473–496. https://doi.org/10.1113/jphysiol.1952.sp004718
(45) Hodgkin, A. L.; Huxley, A. F. The Dual Effect of Membrane Potential on Sodium Conductance in the Giant Axon of Loligo. *The Journal of Physiology* 1952, *116* (4), 497–506. https://doi.org/10.1113/jphysiol.1952.sp004719
(46) The Nobel Prize in Physiology or Medicine 1963. NobelPrize.org. https://www.nobelprize.org/prizes/medicine/1963/summary
(47) The Nobel Prize in Physiology or Medicine 1991. NobelPrize.org. https://www.nobelprize.org/prizes/medicine/1991/summary
(48) Tam, D. C. Interspike Interval Decoding Neural Network. Patent No. US-5216752-A, June 1, 1993. https://image-ppubs.uspto.gov/dirsearch-public/print/downloadPdf/5216752
(49) Tam, D. C.; Perkel, D. H. A Model for Temporal Correlation of Biological Neuronal Spike Trains; 1989. https://doi.org/10.1109/IJCNN.1989.118667
(50) Tam, D. C. Decoding of Firing Intervals in a Temporal-Coded Spike Train Using a Topographically Mapped Neural Network. In *IJCNN. International Joint Conference on Neural Networks*; 1990. https://doi.org/10.1109/IJCNN.1990.137965
(51) Tam, D. C. A Cross-Interval Spike Train Analysis: The Correlation between Spike Generation and Temporal Integration of Doublets. *Biological Cybernetics* 1998, *78*, 95–106. https://doi.org/10.1007/s004220050417
(52) Fitzurka, M. A.; Tam, D. C. A Joint Interspike Interval Difference Stochastic Spike Train Analysis: Detecting Local Trends in the Temporal Firing Patterns of Single Neurons. *Biological Cybernetics* 1999, *80* (5), 309–326. https://doi.org/10.1007/s004220050528
(53) Tam, D. C.; Perkel, D. H.; Tucker, W. S. Temporal Correction of Multiple Neuronal Spike Trains Using the Back-Propagation Error Correction Algorithm. *Neural Networks* 1988. https://doi.org/10.1016/0893-6080(88)90312-7
(54) Tam, D. C. A Spike Train Analysis for Detecting Temporal Integration in Neurons. *Neurocomputing* 1999. https://doi.org/10.1016/S0925-2312(99)00104-6

(55) Lemaréchal, C. Cauchy and the Gradient Method. *Documenta Mathematica* 2012, *251* (254), 10. https://content.ems.press/assets/public/full-texts/books/251/chapters/online-pdf/978-3-98547-540-7-chapter-4938.pdf

(56) Rumelhart, D. E.; Hinton, G. E.; Williams, R. J. Learning Representations by Back-Propagating Errors. *Nature* 1986, *323* (6088), 533–536. https://www.nature.com/articles/323533a0

(57) Rumelhart, D. E.; McClelland, J. L.; Group, P. R. *Parallel Distributed Processing, Volume 1: Explorations in the Microstructure of Cognition: Foundations*; The MIT Press, 1986. https://doi.org/10.7551/mitpress/5236.001.0001

(58) Amari, S. A Theory of Adaptive Pattern Classifiers. *IEEE Transactions on Electronic Computers* 2006 (3), 299–307. https://doi.org/10.1109/PGEC.1967.264666

(59) Robbins, H.; Monro, S. A Stochastic Approximation Method. *The Annals of Mathematical Statistics* 1951, 400–407. https://www.jstor.org/stable/2236626

(60) The Nobel Prize in Physics 2024. NobelPrize.org. https://www.nobelprize.org/prizes/physics/2024/hinton/facts

APPENDIX A

List of Top Forensic Genetics Companies That Provide Services for Families and Law Enforcement Agencies

Company	URL	Description of Services
1. Bode Technology	http://www.bodetech.com/	Offers Forensic Investigative Genetic Genealogy (FIGG), combining DNA technology and genealogical research to generate actionable leads. The firm provides services for solving violent crimes and identifying human remains. Bode has also been a leading provider of cutting-edge DNA collection products. With resources for references, crime scene evidence, and advanced storage solutions.

(*Continued*)

Company	URL	Description of Services
2. Othram	http://www.othram.com/	Specializes in next-generation sequencing for forensic evidence, including whole genome sequencing and proprietary KinSNP® analysis. The firm offers in-house processing of evidence, from extraction to genealogy. Othram also features essential infrastructure and processes for testing forensic DNA evidence, including automatic human enrichment, separate unidirectional workflows
3. AdvancedDNA	https://www.advancedna.org/	Provides investigative genetic genealogy services. AdvancedDNA also leverages Investigative Genetic Genealogy to provide DNA-based leads in investigations and identification research. Adoption, Decedent, Living, and Training.
4. DNA Labs International	http://www.dnalabsinternational.com/	Focuses on forensic DNA analysis for law enforcement and government labs.
5. Signature Science	http://www.signaturescience.com/	Supports law enforcement with casework solutions, including traditional and specialized sample analysis.

APPENDIX B

List of Miscellaneous Forensic Genetics Resources

Resource	URL	Description of Resource
1. DNA Resource.com	http://www.DNAResource.com/	A comprehensive resource for information about forensic DNA testing, including STRBase.
2. DNA POLICY - Abington Medical Centre	www.abingtonmedicalcentre.co.uk	A forum for discussing DNA-related policy issues.
3. DNA Litigation Legal Support Page	https://www.dnalegalservices.org/	Provides legal support and resources for cases involving DNA evidence.
4. DNA Labs International	http://www.dnalabsinternational.com/	Focuses on forensic DNA analysis for law enforcement and government labs.
5. FBI's Combined DNA Index System (CODIS)	http://www.oig.justice.gov/	A national DNA database is used for matching DNA profiles from crime scenes.
6. National Clearinghouse for Science, Technology, and the Law (NCSTL)	http://www2.stetson.edu	Offers information on the intersection of science, technology, and law.

(Continued)

Resource	URL	Description of Resource
7. National Forensic Science Technology Center (NFSTC)	https://ignite.fiu.edu/give-now/giving-opportunities/units-and-divisions/research/the-national-forensic-science-technology-center-nfstc-discretionary-fund/index.html	Focuses on advancing forensic science technology and research.
8. National District Attorneys Association (NDAA)	http://ndaa.org	A professional organization for district attorneys, offering resources and training.
9. National Institute of Justice (NIJ)	http://nij.ojp.gov	Provides funding and research for various forensic science areas, including DNA analysis.
10. American Academy of Forensic Sciences	http://www.aafs.org	An organization for forensic professionals, offering certification and resources.
11. American Society of Crime Laboratory Directors (ASCLD)	http://www.ascld.org	A professional organization for crime lab directors, focusing on accreditation and quality assurance.
12. Innocence Network	http://innocencenetwork.org	A network of organizations working to free wrongly convicted individuals, often involving DNA evidence.
13. Forensic Resources	http://forensicresources.org	A website offering articles, reports, and training on various forensic science topics, including DNA.
14. Forensics TTA	http://www.forensicstta.org	A resource for forensic science professionals, offering technical assistance and training.

(*Continued*)

Resource	URL	Description of Resource
15. A Simplified Guide to Forensic Science	http://forensicresources.org	The National Forensic Science Technology Center created this website to explain in simplified terms the principles of each type of forensic analysis and how the analysis is performed. Topics include DNA, digital evidence, fingerprints, firearms, trace evidence, blood stains, and more.
16. Bureau of Alcohol, Tobacco, Firearms, and Explosives	http://www.atf.gov	The federal agency responsible for administering and enforcing the criminal and regulatory provisions of Federal laws pertaining to explosives enforcement and arson.
17. Death by Fire	http://www.pbs.org	This Frontline documentary investigates the case of Cameron Todd Willingham, a man who was executed in Texas in 2004 for allegedly setting the fire that killed his three young daughters. Prior to his execution, reports were submitted to Texas officials raising serious doubts about the science used to determine that the fire was arson.

(*Continued*)

Resource	URL	Description of Resource
18. Fire and Explosion Investigation Subcommittee	http://www.nist.gov	Focuses on standards and guidelines related to the investigation, analysis, and interpretation of crime scenes where arson or use of explosives is suspected. This OSAC has replaced what the Technical Working Group for Fire and Explosions
19. Ignitable Liquids Reference Collection Database	http://ilrc.ucf.edu	The National Center for Forensic Science at the University of Central Florida has tested over 600 commercial products and compiled information about their molecular composition in an online database.
20. International Association of Arson Investigators, Inc.	http://www.firearson.com	An association of more than 5,000 fire investigation professionals. The website has some information about training and standards, but most of the material is available only to members.
21. NC International Association of Arson Investigators, Inc.	http://www.nciaai.com	The NC chapter of the IAAI. Finds information about relevant NC case law, local training, and membership.
22. NIJ Funded Software Tools, Apps, and Databases	http://nij.ojp.gov	Free or low-cost software tools that may be of assistance in understanding forensic evidence disciplines of digital forensics, arson investigation, DNA, death investigation, and more.

(*Continued*)

Resource	URL	Description of Resource
23. The Arson Research Project	http://forensicresources.org	This criminal justice research project has a mission of examining the reliability of evidence used in the investigation and prosecution of arson and identifying convictions obtained based on unreliable evidence.
24. Forensic Ontology—Bitemark Evidence	www.nist.gov	Includes articles and news reports on the controversial area of bite mark analysis, as well as information about cases where bite mark evidence has played a role.
25. Bloodstain Pattern Analysis Video Collection	http://www.ameslab.gov	Over 1,000 video clips were made of blood spatter experiments that are typically performed by individuals attending a basic bloodstain pattern analysis course. The videos were made possible through a National Institute of Justice (NIJ) grant and are available to be viewed or downloaded.
26. ANAB Directory of Accredited Organizations	http://anab.ansi.org	This database can be used to identify which laboratories are accredited.

(*Continued*)

Resource	URL	Description of Resource
27. A Simplified Guide to Forensic Science	http://forensicresources.org	The National Forensic Science Technology Center created this website to explain in simplified terms the principles of each type of forensic analysis and how the analysis is performed. Topics include DNA, digital evidence, fingerprints, firearms, trace evidence, blood stains, and more.
28. Crime scene forensics: How does it work?	http://www.forensicsciencesimplified.org	A series of narrated images that describe various crime scene investigation techniques and laboratory tests.
29. Anatomy of the Human Body	http://Bartleby.com	Bartleby.com edition of Gray™s Anatomy of the Human Body features 1,247 illustrations and a subject index with 13,000 entries.
30. Medicolegal Death Investigation Subcommittee	http://www.nist.gov	The Medicolegal Death Investigation Subcommittee of OSAC focuses on standards and guidelines related to sudden, unnatural, unexplained, or suspicious deaths, including homicides, suicides, unintentional fatal injuries, drug-related deaths, and other deaths that are sudden or unexpected; determination of the cause and manner of death.

(*Continued*)

Resource	URL	Description of Resource
31. N.C. Office of the Chief Medical Examiner	http://www.ocme.dhhs.nc.gov	The North Carolina Medical Examiner System is a network of medical doctors and allied health professionals throughout North Carolina who voluntarily devote their time, energy, and medical expertise to see that deaths of a suspicious, unusual, or unnatural nature are adequately investigated. The OCME investigates all deaths in North Carolina due to injury or violence.
32. NIJ Funded Software Tools, Apps, and Databases	http://nij.ojp.gov	Free or low-cost software tools that may be of assistance in understanding forensic evidence disciplines of digital forensics, arson investigation, DNA, death investigation, and more.
33. Sample Cross-Examination Transcripts	http://forensicresources.org	Sample direct and cross-examinations of various forensic witnesses, including a firearm/toolmark expert, fingerprint expert, pathologist, DNA expert, and other forensic experts.

(*Continued*)

Resource	URL	Description of Resource
34. Evolutions in DNA Forensics	http://www.llrx.com	Law and Technology Resources for Legal Professionals compiled this list of books, articles, and other publications that treat different aspects of DNA evidence, including DNA databases, forensic interpretation of DNA evidence, 4th and 5th Amendment concerns, and more.
35. Forensic Bioinformatics	http://www.bioforensics.com	DNA experts in OH with videos, articles, and a glossary of DNA testing terms available for free download on their website.
36. Scientific Working Group on DNA Analysis Methods (SWGDAM)	http://www.swgdam.org	A group of approximately 100 scientists representing federal, state, and local forensic DNA laboratories in the United States and Canada. SWGDAM periodically issues new Interpretation Guidelines that crime labs then implement.
37. Short Tandem Repeat DNA Internet Database	http://strbase-archive.nist.gov	Forensic DNA analysis resources compiled by John Butler. Includes training materials, technical information, and a collection of data by the National Institute of Standards and Technology (NIST).
38. Supreme Court Upholds Taking DNA Upon Arrest	http://www.ojp.gov	Supreme Court case Maryland v. King, which allows the taking of DNA from arrestees.

APPENDIX C

Frequently Asked Questions

Question	URL	Partial Answer
1. How is genetic genealogy used in a forensic context?	https://www.justice.gov	Forensic genealogy is law enforcement's use of DNA analysis combined with traditional genealogy research to generate investigative leads for unsolved violent crimes.
2. How does Figg work?	https://www.police.nsw.gov.au	FIGG combines traditional genealogy using historical and government records along with open-source intelligence to identify living individuals. Family trees are built back from matches to the unknown individual to identify the most recent common ancestor between two matches.
3. What type of DNA does a genetic genealogy need?	https://guides.loc.gov	Autosomal DNA (atDNA) is best to help identify ancestors within the most recent 5-7 generations of a family tree.
4. What is the investigative genetic genealogy method?	https://www.tdcaa.com	IGG is an innovative investigative approach that combines traditional genealogy with advanced DNA analysis to solve crimes. The technique is primarily used on cold cases, but the future of many criminal investigations that are not initially solved may be based in the world of IGG.

(*Continued*)

Question	URL	Partial Answer
5. How does forensic genealogy solve crimes?	https://bjs.ojp.gov	Forensic Genealogy utilizes traditional genealogy research with advanced DNA testing to help identify potential links to unknown profiles. Intermountain Forensics is a nonprofit Forensic DNA Testing Laboratory employing cutting-edge molecular biology to cold cases.
6. How does DNA play a role in forensics?	https://bjs.ojp.gov	DNA is responsible for most inherited traits in humans. forensic scientists use DNA from blood, semen, skin, saliva, or hair recovered from crime scenes to identify possible suspects through DNA profiling, during which the length of repetitive DNA sections is compared.
7. Is genetic genealogy legal?	https://jolt.richmond.edu	There is no national standard for the use of genetic genealogy. Twelve states allow for the use of it, while Montana and Maryland are the only two states that have explicit statutory limitations on the use of genetic genealogy and familial searches.
8. What does FIGG stand for?	https://verogen.com	Forensic investigative genetic genealogy (FIGG) is a powerful tool for solving violent crimes such as homicides or sexual assaults, and identifying human remains from missing persons investigations and conflict zones.

(*Continued*)

Question	URL	Partial Answer
9. How accurate is genetic genealogy?	https://www.ancestry.com	Accuracy is very high when it comes to reading each of the hundreds of thousands of positions (or markers) in your DNA. With current technology, AncestryDNA ® has, on average, an accuracy rate of over 99 percent for each marker tested.
10. Do siblings have the same DNA genealogy?	https://sanogenetics.com	The differences in genetic code between siblings are due to variations in the chromosomes passed down to them from their parents. This, more often than skeletons in the closet, also explains why some siblings will receive different ethnicity or ancestry results.
11. How is genealogy used in forensics?	https://pmc.ncbi.nlm.nih.gov	Forensic Genealogy typically refers to estate and probate cases to identify/find heirs, the identification of living descendants of fallen soldiers for their repatriation, and other historical investigations.

APPENDIX D

Cold Case Studies

Cold Case	URL	Description
1. The Zodiac Killer	https://www.fbi.gov	A serial killer who terrorized California in the late 1960s and early 1970s, sending letters and ciphers to newspapers.
2. The Black Dahlia	https://www.fbi.gov	The 1947 murder of Elizabeth Short in Los Angeles, a case that has captivated the public's imagination for decades.
3. The Hall-Mills Murders	https://www.mrlocalhistory.org	A series of murders in the 1960s and 1970s, with two women and three children found dead in rural locations.
4. JonBenét Ramsey	https://bouldercolorado.gov	The 1996 murder of a young girl in Colorado, a case that remains unsolved despite extensive media coverage and investigations.
5. Lizzie Borden	https://lizzie-borden.com	The 1892 axe murders of her father and stepmother, with Lizzie Borden being the main suspect, but she was never formally convicted.
6. The D.B. Cooper Hijacking	https://www.fbi.gov	A 1971 skyjacking where a man parachuted out of an airplane with a ransom, his fate and the whereabouts of the money remain unsolved.

(*Continued*)

Cold Case	URL	Description
7. The Tylenol Tamperings	https://www.ebsco.com	The 1982 incident where seven people in Chicago died after taking tainted Tylenol capsules, the culprit(s) were never apprehended.
8. The Anthrax Scare	https://www.fbi.gov	A series of anthrax-related mailings in 2001 caused widespread fear and panic; the perpetrator(s) were never caught.
9. The Boy in the Box	https://www.nytimes.com	A child found dead in a cardboard box in 1980, his identity was never determined.
10. Alcatraz Escape	https://www.fbi.gov	Alcatraz Escape. The fate of three men, Frank Morris, John Anglin, and his brother Clarence Anglin, who made a daring escape.
11. Colonial Parkway murders	https://www.fbi.gov	The Colonial Parkway murders were the serial murders of at least ten people in the U.S. state of Virginia between 1986 and 1989.
12. Jack the Ripper	https://www.jack-the-ripper.org	Jack the Ripper was an unidentified serial killer who was active in and around the impoverished Whitechapel district of London, England, in 1888.
13. Cameo murders	https://www.cameo.com	Despite two trials and two appeals, to this day the Cameo murders remain one of the great unsolved murder cases of the 20th century.

(*Continued*)

Cold Case	URL	Description
14. Tupac Shakur	https://www.rollingstone.com	Tupac Shakur's unsolved murder investigation, the "California Love" rapper's drive-by shooting, and the ensuing trial.
15. Jimmy Hoffa	https://www.ebsco.com	Hoffa, a prominent labor leader who served as president of the Teamsters union from 1957 to 1967, vanished on July 30, 1975, under mysterious circumstances.
16. Lady in the box	https://open.spotify.com	The murder of an Ohio woman in 1974
17. Mary Rogers	https://www.historicalcrimedetective.com	Mary Cecilia Rogers was an American murder victim whose story became a national sensation (1841).

Glossary

Census Records An official survey of the population of a country, state, region, or locality that is carried out in order to find out how many people reside there and to obtain details of such things as people's ages and jobs, family relations, and location of birth.

cM Centimorgan. A centimorgan is a unit of shared genetic linkage between persons.

CODIS (Combined DNA Index System) A national DNA database that stores DNA profiles from crime scenes and convicted offenders. Also, a national database of DNA profiles that law enforcement agencies use to compare crime scene DNA with existing profiles.

Direct-to-Consumer DNA Testing Companies that offer DNA testing to individuals for genealogy and other purposes, often contributing to genealogical databases. Also, services that allow individuals to test their DNA and upload their results to public databases.

DNA Deoxyribonucleic acid.

DNA Profile A unique pattern of DNA markers that is characteristic of an individual.

DNA Profiling The process of analyzing DNA to identify individuals or link them to crime scenes

DNA Raw Data file Genetic information that has been digitized as computer data.

DOJ Department of Justice.

Family Tree A visual representation of an individual's family history, used to trace familial connections in FIGG.

Family Tree Building The process of constructing a family tree based on DNA matches and genealogical records to identify the target.

FIGG Forensic Investigative Genetic Genealogy. This investigative technique will also have abbreviations of IGG (Investigative Genetic Genealogy), FGG (Forensic Genetic Genealogy), and IG (Investigative Genealogy).

Forensic Genetic Genealogical DNA Analysis (FGG) The specific DNA analysis used in FIGG focuses on SNPs across the entire genome.

Genomic databases Databases that contain DNA profiles and genealogical information from individuals who have participated in DNA testing.

Genealogical Databases Online databases where individuals have uploaded their DNA profiles for comparison and analysis.

Genetic Relatives Individuals who share a substantial amount of DNA with a target individual.

Historical Records Documents like birth certificates, census records, and marriage records are used in genealogical research.

Informed Consent This is the practice of informing and educating a participant in an investigation of all the information they may need to make a decision on whether they wish to participate in the investigation or not. In FIGG cases this is usually utilized when searching consumer DNA Databases for shared persons and when asking reference testers to volunteer their DNA.

Investigative Lead This is information in a criminal investigation that may assist in finding building blocks to identify the perpetrator of the crime, or UHR. This often includes how the crime was committed, why the crime was committed, and potentially who committed the crime.

Mitochondrial DNA (mtDNA) DNA found in the mitochondria of a cell, which can be used to link individuals through maternal lineages.

Nuclear DNA DNA found in the nucleus of a cell, which is the focus of most forensic DNA analysis.

Obituary a notice of a person's death most commonly found in newspapers or online that may contain stories about a person's life, as well as names of family and friends, and when and where upcoming memorial services will be.

Partial DNA Profile A DNA profile that does not contain results for all 13 core loci (locations on DNA).

Putative Perpetrator Refers to the suspect or person of interest in a criminal investigation. Often associated with FIGG cases where suspect DNA evidence is left on the crime scene.

Random-Match Probability The statistical chance that a DNA profile could belong to someone other than the individual to whom it is attributed.

Reference Samples DNA samples collected from known individuals to compare with the DNA found at a crime scene.

Shared DNA Blocks Sections of DNA that individuals share with their relatives, which are used to build family trees in FIGG.

SNP Single-nucleotide polymorphism. Small variations in DNA sequences that are used in FGG analysis to identify shared DNA blocks between individuals.

SNP (Single Nucleotide Polymorphism) A variation in the DNA sequence that is used in FIGG to identify shared DNA blocks between individuals.

STR Short tandem repeats.

STR DNA (Short Tandem Repeats) A type of DNA marker used in traditional forensic DNA analysis.

STR DNA Typing A traditional forensic DNA analysis method that focuses on Short Tandem Repeats (STRs), which are specific DNA sequences.

Target The unknown individual being investigated in a FIGG case (a suspect or unidentified remains).

UHR Unidentified Human Remains.

X-DNA DNA from the X chromosome.

Y-DNA DNA from the Y chromosome.

Index

Pages in **bold** refer to tables.

D

E

For Product Safety Concerns and Information please contact our EU representative GPSR@taylorandfrancis.com
Taylor & Francis Verlag GmbH, Kaufingerstraße 24, 80331 München, Germany

www.ingramcontent.com/pod-product-compliance
Ingram Content Group UK Ltd.
Pitfield, Milton Keynes, MK11 3LW, UK
UKHW022210190926
475702UK00003B/283
* 9 7 8 1 0 3 2 7 1 4 9 7 4 *